Reproductive Medicine for Clinicians

Volume 3

This series will focuses on and presents developments in knowledge and practice within all aspects of reproductive medicine.
It will help to cover the important gap between the new possibilities offered by the most recent investigations and technical developments and the application in clinical practice.
The series will be a useful tool for professionals and practitioners in the fields of Gynecology, Obstetrics, and Human Reproduction. Trainees interested in the most complete information on the developments of reproductive medicine will benefit as well.
This series is published in partnership with the International Academy of Human Reproduction (IAHR)

Joseph G. Schenker • Martin H. Birkhaeuser
Andrea R. Genazzani • Liselotte Mettler
John J. Sciarra
Editors

Hot Topics in Human Reproduction

Ethics, Law and Society

The International Academy of Human Reproduction

Springer

Editors
Joseph G. Schenker
Department of Obstetrics and Gynecology
Hebrew University - Hadassah
Medical Center
Jerusalem, Israel

Martin H. Birkhaeuser
Gynaecological Endocrinology and
Reproduction
University Hospital of Bern
Bern, Switzerland

Andrea R. Genazzani
Department of Reproductive Medicine
University of Pisa
Pisa, Italy

Liselotte Mettler
Campus Kiel
University Clinics of Schleswig-Holstein
Kiel, Germany

John J. Sciarra
Department of Obstetrics and Gynecology
Northwestern University Medical School
Chicago, IL, USA

ISSN 2523-3599 ISSN 2523-3602 (electronic)
Reproductive Medicine for Clinicians
ISBN 978-3-031-24902-0 ISBN 978-3-031-24903-7 (eBook)
https://doi.org/10.1007/978-3-031-24903-7

This Springer imprint is published by the registered company Springer Nature Switzerland AG
The registered company address is: Gewerbestrasse 11, 6330 Cham, Switzerland

Preface

The International Academy of Human Reproduction is delighted with the publication of the third volume of the series titled Reproductive Medicine for Clinicians, which is published by the loyal partner Springer.

The present volume focuses on hot topics in human reproduction: ethics, law, and society.

The contents include original articles, reviews, and views arranged in five sections:

1. Covid-19 Pandemic
2. Beginning of Human Life
3. Rights to Reproduce
4. Innovative Technologies
5. Some Difficult Patients

The chapters are written by established pioneers and experts in human reproduction, and it is with great appreciation and gratitude that we thank them for their enormous contribution to this volume.

The main objectives of the Academy are to extend the knowledge in all clinical aspects of human reproduction, to encourage clinical experience and promote scientific thoughts and investigation, and to consider the ethical and social implications of current practice of human reproduction.

The fellows of the Academy are elected based on their significant contribution to the field and must be acknowledged as world leaders in the discipline. The fellows of the Academy are selected from among applicants from the fields of clinical medicine, medical and biological sciences, and others related to reproductive health.

Starting in 1974 in Rio de Janeiro, the Academy has held successful congresses every 3 years in Europe, Asia, Africa, North and South America, and Australia.

Due to Covid-19 pandemic, we have continued to exchange knowledge and research among members and the human reproductive community by webinars and congress in Columbia.

We regret that we were unable to hold the 19th Congress, Jerusalem 2021, which was postponed (Venice, March 2023).

The series Reproductive Medicine for Clinicians is a useful tool for professionals and practitioners in the fields of gynecology, obstetrics, and human reproduction. Trainees interested in the most complete information on the developments of reproductive medicine will benefit as well.

On behalf of the International Academy of Human Reproduction (IAHR), I trust that you will support the suitability of this high-quality book series to human reproduction.

Professor Joseph G. Schenker, MD, FRCOG, FACOG
President, The International Academy of Human Reproduction
Jerusalem, Israel

Contents

Part I

Covid-19 Pandemic

COVID-19: Pandemic Effect on Human Reproduction

1

Yaakov Bentov and Joseph G. Schenker

Introduction

Since its emergence as a cluster of severe viral pneumonia identified in Wuhan, China, at the end of 2019, the coronavirus disease-19 (COVID-19) had rapidly spread to become a global pandemic affecting, at the time of writing, close to half a billion people and resulting in 18.2 million fatalities worldwide [1]. This pandemic was the first in the era of modern medicine. It was also the first to be managed with a global coordinated response that included the development and introduction of multiple vaccines and specific treatments shortly after its eruption. Also, unlike the previous pandemic that occurred a century ago, the "Spanish Flu," this pandemic was faced with a popular movement of scientific denialism fueled by conspiracy theories and fractured sources of information [2]. COVID-19 is a multisystemic disease caused by a novel virus, severe acute respiratory syndrome coronavirus 2 (SARS-CoV-2), which has short-term effects caused by temporary loss of function as well as long-term effects caused by tissue damage. These may involve any organ system [3, 4]. Quite early during the pandemic, it was recognized that although most infected people will have either no or mild symptoms, there are several populations at risk for severe morbidity and mortality, among them pregnant women. Despite similar rates of contamination, pregnant women present with a higher rate of hospitalization (31.5% vs 5.8%), a higher adjusted risk for ICU admission [aRR 1.5, 95% confidence interval (CI): 1.2–1.8], a higher risk to receive mechanical ventilation [aRR 1.7, 95% confidence interval (CI): 1.2–2.4], and extracorporeal membrane oxygenation (ECMO) [aRR 2.4, 95% confidence interval (CI): 1.5–4.0]

Y. Bentov · J. G. Schenker (✉)
Department of Obstetrics and Gynecology, Hebrew University-Hadassah Medical Centre, Jerusalem, Israel
e-mail: joseph.schenker@mail.huji.ac.il

J. G. Schenker et al. (eds.), *Hot Topics in Human Reproduction*, Reproductive Medicine for Clinicians 3, https://doi.org/10.1007/978-3-031-24903-7_1

compared to their age-matched counterparts [5]. Together with apprehension of potential detrimental effects of COVID-19 and its vaccine on male and female reproductive function, this has led to an unprecedented public interest in this topic.

The SARS-CoV-2 is a coronavirus (CoV) that is found globally in many animal species. It is part of the "*Orthocoronaviridae*" subfamily that are divided into four groups α to δ, with only groups α and β affecting mammals. The highly pathogenic SARS-CoV, MERS-CoV, and SARS-CoV-2 are βCoV. CoV are RNA viruses, enveloped by a host-derived lipid membrane embedded with viral proteins. The proteins protruding from its membrane give these viruses their ultramicroscopic typical halo appearance that gave this group of pathogens their name corona (crown in Latin) [6]. The RNA of the virus is single stranded, and it has the same orientation as mRNA. CoV viruses have the largest genome of all RNA viruses. Due to the high rate of mutations related to its unique RNA-dependent RNA polymerase and homologous recombination, coronaviruses acquired a great diversity that enabled these viruses to infect many species and allow inter-species transmission [7]. Genomic sequencing of the virus showed it to be related to two other human CoVs: severe acute respiratory syndrome coronavirus 1 (SARS-CoV-1) and Middle East respiratory syndrome-related coronavirus (MERS-CoV) as well as to a bat CoV RaTG13.

The CoV RNA encodes four essential structural proteins including the nucleocapsid protein that surrounds the viral genome and three membrane proteins: the S-glycoprotein (spike protein), the matrix (M) protein, and the envelope (E) protein. The "spike protein" (S), a transmembrane glycoprotein of the CoV, has two subunits: S1 responsible for the attachment to the host cell receptor and S2 that allows virus-host membrane fusion. The virus attaches to the angiotensin-converting enzyme 2 (ACE2), a receptor found mostly on surfaces of human respiratory cells and uses it as a point of entry [8, 9]. The ACE2 receptor serves as a critical port of entry to SARS-CoV and SARS-CoV-2 but not MERS-CoV [10]. The S1 protein has a high affinity to the ACE2 receptor; however, for fusion to occur, the virus needs to shed the S1 protein and activate S2. The cleavage of the covalent bond between S1 and S2 occurs after the assembly of the virus inside the host cell via the action of a Furin protease (Fig. 1.1). The cleavage of this covalent bond allows for an easier shading of S1 subunit and is a prerequisite for the activation of the S2 subunit. The next stage involves the activation of S2 subunit by cleavage of the S2′ site by the action of the co-expressed transmembrane serine protease-2 (TMPRSS2) present at the cell surface. The activation of the S2 component initiates membrane fusion by creating fusion pores that enable the insertion of the viral genome into the cell. Membrane fusion leads to endocytosis of the virus into the cell and into its nucleus to start replication. The newly formed viral DNA is used to form the viral proteins which are then packaged, transferred to the cell membrane, and released to infect other cells (Fig. 1.1) [11]. It is therefore imperative for tissues to co-express the ACE2 receptor and TMPRSS2 genes to become targets for the SARS-CoV-2. Co-expression of ACE2 and TMPRSS2 is only present on the lung, large and small intestine, esophagus, brain, heart, kidney, testis, and fallopian tubes [12]. Although there is a clear preference of the ACE2-TMPRSS2 mode of host cell entry, an alternative route of host cell entry in cells devoid of TMPRSS2 has been described. This

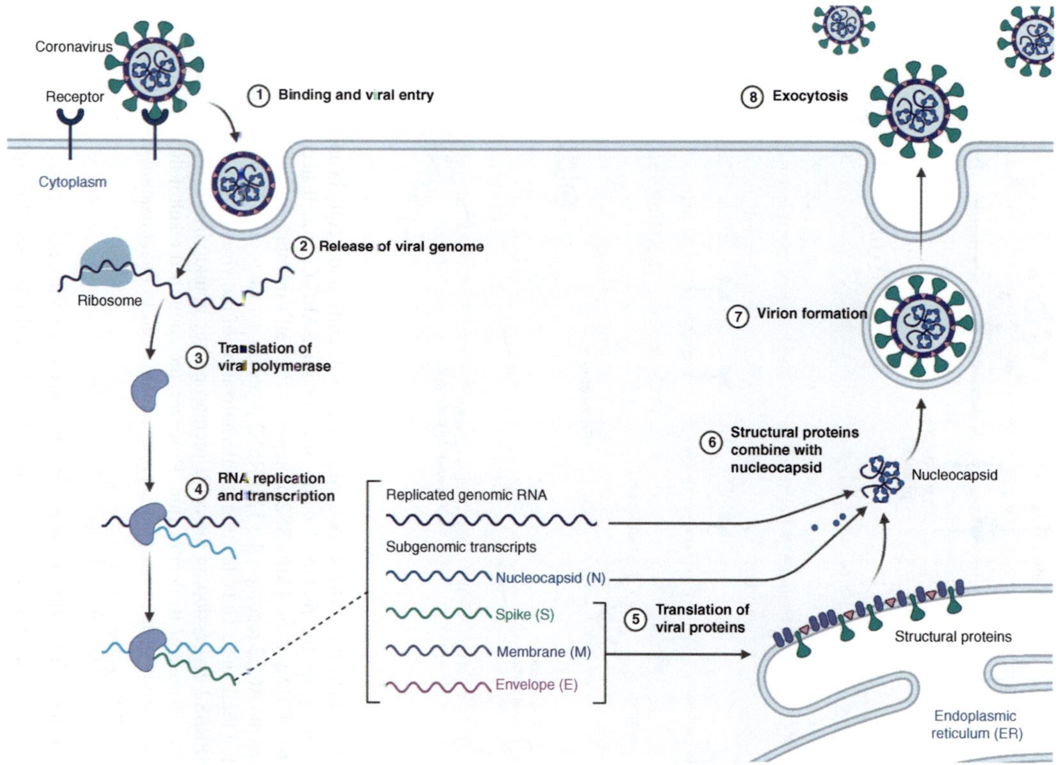

Fig. 1.1 The cycle of the SARS-CoV-2 virus starts with binding to the ACE2 receptor, followed by membrane fusion and internalization of the virus into the cell and release of the viral RNA. The viral RNA then using the cell's machinery replicates its RNA and translates it into the four viral proteins: the nucleocapsid, spike, membrane, and envelope proteins. These components are than assembled to create the virion which is then released from the cell in a process of exocytosis to infect other cells

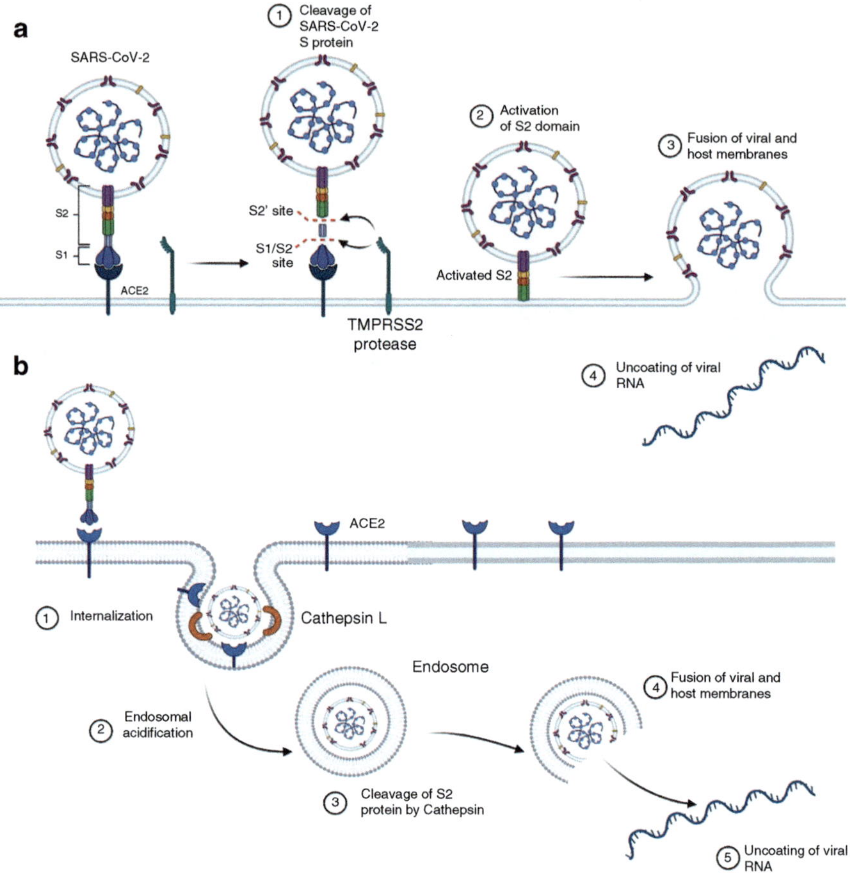

Fig. 1.2 An infection by the SARS-CoV-2 virus follows its entry into a cell via one of two mechanisms. (**a**) This is by far the preferred mode of entry for SARS-CoV-2. It necessitates the co-expression of the ACE2 and the TMPRSS2 protease on the cell's membrane. Following binding of the S1 subunit to the ACE2 receptor, the TMPRSS2 protease activates the S2 subunit that acts as a membrane fusion facilitator. Thus, leading to internalization of the virus into to the cell. (**b**) In the absence of TMPRSS2 expression on the cell's membrane, the bound virus is internalized into an endosome for the purpose of its degradation. However, the combined result if the acidification of the endosome and action of the cathepsin L protease may lead to the activation of the S2 protein and membrane fusion

path involves internalization of the ACE2-bound virus via endocytosis. There, in the endolysosome, the S2′ is cleaved by the action of cathepsins, and the S2 subunit is activated (Fig. 1.2). However, the limited effect of hydroxychloroquine, a known inhibitor of endosomal acidification, suggests that this is not the main mode of entry used by the SARS-CoV-2 virus [13]. The ACE2 is a key element in counterbalancing the action of the renin–angiotensin–aldosterone system (RAAS). The RAAS system is activated to compensate for low blood pressure by activation of the angiotensin receptor 1 (ATR1) that leads to vasoconstriction and increased absorption of sodium and water but also to cell proliferation, inflammation, and fibrosis. ACE2 acts through activation of its mediators, angiotensin 1–7 [ANG-(1–7)] and the major receptor mitochondrial assembly 1 (MAS1). Their activation leads to vasodilatation and has prevention of inflammation, fibrosis, and cell proliferation thus balancing overactivation of RAAS. SARS-CoV-2 may inhibit the protective action of ACE2 thereby leading to dysregulation and overactivation of RAAS leading to an increased expression of membranous ACE2 that serves as a port of entry to other SARS-CoV-2 viruses (Fig. 1.3) [10]. Studies had also reported on a wide expression of ACE2 in human placenta and its vasculature that peak in early gestation [14, 15]. Thus, potentially both male and female reproductive systems seem to be susceptible to SARS-CoV-2 infection and a short-term or long-term dysfunction. Also, common symptoms associated with the acute infection, such as fever and

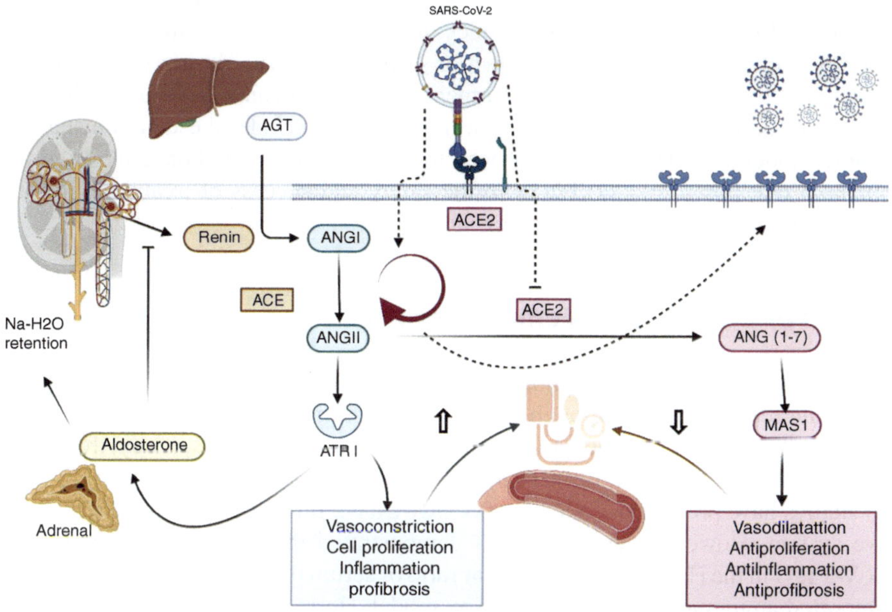

Fig. 1.3 The renin–angiotensin–aldosterone system (RAAS) is a short circuit regulatory system that responds to changes in blood pressure and solute composition at the juxtaglomerular apparatus in the cortex of the kidney. Activation of the system leads to secretion of renal renin, an enzyme that cleaves the constantly produced hepatic angiotensinogen (AGT) to angiotensin I (AGT I). The second phase of the activation of this hormone is facilitated by the action of the angiotensin-converting enzyme (ACE) that creates its active form, angiotensin II (AGT II). AGT II following binding to its receptor (ATRI) leads to vasoconstriction as well as to the adrenal secretion of the mineralocorticoid, Aldosterone that in turn promotes renal absorption of sodium and water. These actions that lead to an increase in blood pressure, simultaneously activate a counter-regulatory mechanism to prevent an over response. This involves the activation of the ACE2 receptor that via its mediators, the angiotensin 1–7 and the mitochondrial assembly system 1 promote vasodilatation as well as anti-inflammatory and anti-proliferative actions. The SARS-CoV-2 virus by binding and inactivating the ACE2 leads to over net effect of RAAS and in response an increased expression of membranous ACE2 that promotes further viral cell entry

hypercoagulability, may affect both male and female reproduction. In addition, viral proteins show similarity to placental proteins and could, in theory, interfere with placenta formation [16, 17]. Patients undergoing fertility treatment are in even more complicated circumstances; as they cannot conceive naturally, their future pregnancy may be postponed until the risk of infection declines, while attempts to conceive during the pandemic put them in a potentially higher risk of infection due to frequent visits to fertility clinics or hospitals in which the likelihood of exposure to the virus is higher. Furthermore, some of the reported risk factors among pregnant women for severe COVID-19-related complications were age over 25 years, pre-pregnancy obesity, chronic hypertension, and pre-pregnancy diabetes [18]. All these risk factors are significantly more common among patients conceiving following in vitro fertilization (IVF) treatment as opposed to spontaneous conceptions, suggesting that pregnant women following IVF may be at a particularly elevated risk for COVID-19-related complications [19]. Worldwide, there were different policies with regard to the activity of IVF clinics during the pandemic. Although in some countries, clinics needed to scale down or stop their activity during waves of high infectivity, for the most part, IVF treatment cycles were conducted during the pandemic excluding patients with a current infection [20]. Under these circumstances, couples undergoing IVF treatments need to cope with the added uncertainty of the potential risk posed by COVID-19 to the safety and success of treatment and the possible implications in case of a pregnancy [21].

Effects of COVID-19 on Female Fertility

Several studies examined the short-term effects of COVID-19 on different aspects of female reproduction. Although ACE2 is expressed in the human ovary, whether this virus binds to ACE-2 receptors in the ovary and which effects, if any, this infection would have on ovarian function, and oocyte quality remains unclear. To date, no studies have presented evidence of SARS-CoV-2 infecting the female reproductive system. However, several studies demonstrated the presence of anti-SARS-CoV-2 IgG in the FF. The reported linear ratio of serum to FF antibody concentration supported an unregulated serum filtration model. Herrero et al. reported on a negative correlation between FF anti-SARS-CoV-2 IgG titer and oocyte and mature oocyte yield and a positive correlation with time interval from infection [22, 23].

Follicular Fluid (FF)

The follicular fluid (FF) is a complex mixture of hormones, cytokines, metabolites, and other proteins that originate from serum filtration as well as granulosa cell secretions. It represents the microenvironment of the oocyte, and its composition has been associated with its quality. Several studies compared FF composition in SARS-CoV-2 recoverees to non-exposed IVF patients. Heparan-sulfate-proteoglycan-2 (HSPG2) is the main estrogen-binding protein in FF and was found to be the FF protein with the highest predictive value for oocyte fertilization and the resulting embryo implantation. Comparison of HSPG2 FF concentration between recent SARS-CoV-2 recoverees [98.14 days from recovery to sampling (range 48–169 days)] to non-exposed showed no difference [23]. Another study showed a lower concentration of IL-1β and VEGF in FF from COVID-19 recoverees. In vitro exposure of granulosa cells to this FF was associated with markers of DNA damage [22].

Oocyte Yield

Several studies compared oocyte yield in COVID-19 recoverees. All the reports show the total number of retrieved oocytes as well as the number of mature oocytes to be unaffected by recent exposure to SARS-CoV-2. The study by Youngster et al. found oocyte yield to be lower in the group of COVID-19 recoverees with exposure to IVF treatment interval of 6 months or longer [22–25].

Steroidogenesis

Other than oocyte maturation, the ovaries are the source of the female sex steroids. The production of estradiol during the follicular phase of the cycle serves several essential physiological roles as well as a marker of the adequacy of ovarian response. Several recent publication studies compared peak serum estradiol as well as FF estradiol of COVID-19 recoverees and non-exposed controls during IVF treatment, all reporting no difference [22–24].

Fertilization Rate and Embryo Quality

The rate of oocyte fertilization is determined by oocyte and sperm function and the quality of their genetic material. Unlike fertilization by intracytoplasmic sperm injection (ICSI), oocyte fertilization by standard IVF is also affected by sperm parameters such as motility, Zona receptor binding, capacitation, and acrosomal reaction. A study that compared oocyte fertilization by either standard IVF or ICSI reported on similar rates in COVID-19 recoverees and non-exposed controls [24]. A comparison of embryo quality as determined by morphological grading was reported by two publications. No measurable difference could be demonstrated between recent COVID-19 recoverees and non-exposed controls [23, 24].

Ovarian Reserve

Serum anti-Mullerian hormone is regarded as the most reliable measure of the residual ovarian reserve, as it is a product of ovarian follicles, not operator dependent, and shows little intra- or inter-cycle variability. Several studies compared serum AMH between COVID-19 recoverees and non-exposed controls. These cohort studies compared women recovering from COVID-19 in different levels of severity.

Three of the four studies showed no detectable difference in AMH, and one study reported a lower AMH among the COVID-19 recoverees. Of note, a much larger cohort study that examined early follicular FSH as a marker of ovarian reserve showed a higher FSH among recoverees, suggesting a lower functional ovarian reserve [26].

Menstrual Cycle Disturbances

Carp-Veliscu et al. recently reviewed publications reporting on menstrual disturbances post COVID-19. The review included 11 recently published studies. Most of these studies were questionnaire-based retrospective cohort studies, and results were inconsistent. Reports ranged from no effect to over 80% reporting on menstrual cycle either shorter or longer as well as changes in flow and dysmenorrhea. Some of the studies associated menstrual cycle changes with the severity of COVID-19 symptoms or levels of stress [26].

Effects on Male Reproduction

Concerns over the impact of COVID-19 on male fertility span over three main questions: (1) Does SARS-CoV-2 infest the testis? (2) Are there short-/long-term detrimental effects of COVID-19 on sperm quality? (3) Can COVID-19 be sexually transmitted via infected sperm?

Despite intense research since the outbreak of the pandemic, there is still no clear answer to these questions [27]. To date, 27 viruses have been detected in human semen in association with viremia. It has been speculated that the presence of viruses in semen may be more common than appreciated and that traditional non-sexually transmitted viruses may be present in the genital secretions. As mentioned earlier, both the ACE2 receptor and TMPRSS2 are expressed in the male reproductive system. There are also multiple reports of detection of the SARS-CoV-2 virus or the spike protein in post-mortem testicular biopsies and semen from COVID-19 patients. There had also been reports of orchitis detected in an autopsy of a COVID-19 patient as well as observations of significant infiltration of immune cells in tests of COVID-19 patients. Yet in most of these reports, contamination of the sample could not be ruled out. Pathological examination of testicleless of man infected with SARS-CoV-2 had also demonstrated changes in the seminiferous tubules, damaged Sertoli cells, and reduction in the number of Leydig cells together with peritubular membrane thickening, fibrosis, and immune cell infiltration. These changes were often associated with low serum testosterone despite relatively elevated LH and FSH. Although these reports may suggest a direct detrimental effect of the virus on testicular function, these changes may also be attributed to the effect of fever commonly experienced by COVID-19 patients. In fact, comparison of sperm parameters between fever-positive and fever-negative male COVID-19 patients showed a significantly lower volume, concentration, and total motility in the febrile group [27]. With regard to the question on SARS-CoV-2 transmission via sperm, there have two recently published reports, the first analyzing semen samples within 24 h of the positive nasopharyngeal swab. This study showed 1/32 samples to be positive for SARS-CoV-2 RNA, however commenting that oral contamination during sample production could not be ruled out [28]. A second report analyzed semen samples from recovered men that were obtained 11–64 days after testing positive for SAR-CoV-2 infection. In this study no viral RNA was detected in any of the samples [29].

IVF Outcome

Several studies compared the outcome of IVF treatment between COVID-19 recoverees and non-exposed patients (reviewed by [24, 26]). These studies were either case-control or cross-sectional cohort studies. Although none of the studies demonstrated a decline in oocyte or mature oocyte yield, some of these studies reported on a decline in embryo quality as defined by either embryo morphological assessment or rate of euploid embryos. However, the reported rates of fertilization, number of cryopreserved embryos, and clinical pregnancy rate were similar [24]. A recent study reported on the outcome of IVF treatments in the county of Lombardy, Italy [30]. Lombardy was one of the areas hardest hit in the early stages of the pandemic. The authors compared several outcome parameters of the pre-exposure cohort, comprised of all the patients undergoing both fresh and frozen IVF treatment cycles before the pandemic (November 2018–March 2019) to the potentially exposed cohort, composed of all IVF cycles conducted during the peak of the COVID-19 outbreak (November 2019–March 2020). Although asymptomatic patients were not tested for SARS-CoV-2, 28% of blood donor samples from that period tested positive for anti-SARS-CoV-2 IgG, suggesting that a similar number of patients undergoing IVF were exposed to the virus. The authors found similar rates of clinical pregnancy, early pregnancy loss, and extrauterine pregnancies [30].

Pregnancy Outcome

A study by Viotti et al. found trophectoderm cells of a day 6 embryo to have the highest co-expression of the ACE-2 and TMPRSS2 and that they are susceptible to the infection through the ACE2 receptor [26]. Therefore, concern over the potential risk to developing pregnancies was expressed. A meta-analysis that studied the incidence of potential vertical transmission reported on 800 newborns testing positive for SARS-CoV-2 out of 308,540 newborns, representing an incidence of 2.6%. However, intrapartum exposure could not be ruled out [31]. A study by Calvo et al. [32] reported on the perinatal outcome of 1347 pregnant women, among them 74 who conceived following IVF, who were infected with SARS-CoV-2 during pregnancy. This multicenter study compared the rate of early pregnancy loss, pregnancy complications, and mode of delivery according to exposure status. They were only able to show a significant increase in the rate of cesarean sections among exposed patients [32]. A meta-analysis of 17 observational studies had also found the rate of early pregnancy loss among COVID-19 patients to be within the expected range

[33]. A recent meta-analysis compared the outcome of over two million births taking place during the pandemic with over 28 million births from the pre-pandemic period. The study showed that pregnancies during the pandemic were associated with a lower rate of spontaneous preterm births (PTB) (nine studies, uaOR 0.91, 95% CI 0.88–0.94) but not in induced PTB (eight studies, uaOR 0.90, 95% CI 0.79–1.01), similar odds for stillbirths (32 studies, uaOR 1.07, 95% CI 0.97–1.18, and 3 studies, aOR 1.18, 95% CI 0.86–1.63), and an increase in average birth weight (nine studies, mean difference 21 g, 95% CI 13–30 g) [34]. A review of several small studies reported on a similar rate of congenital malformations among exposed and exposed pregnancies [35].

The Anti-SARS-CoV-2 Vaccine

Conventional vaccines such as inactivated, live attenuated viruses or their subunits had been at the forefront of humankind's combat against infectious disease. These traditional-type vaccines had proven to be both safe and efficacious, yet their development and production may take many years. With the advent of the COVID-19 global pandemic, there was an urgent need for large-scale development and deployment of a vaccine to halt its rapid expansion. Messenger RNA-based vaccines had been studied since the 1970s. Their non-dependence on animal products or cell culture as well as the rapidness and low cost of their production placed them in the forefront of vaccine research. Yet, implementation of this technology in vaccine development had to await the maturation of a technology that will allow the mRNA segments access into the cell to be translated to the target protein. The development of liposomal nanoparticle carriers in the beginning of the 1990s provided the needed mechanism for cell entry, and by 2017, the stage was ready for the first mRNA anti-rabies vaccine to begin phase I trials. The rapidness of development of the ant-SARS-CoV-2 mRNA vaccines was unprecedented. Only 7 months after the first case of COVID-19 outside of China was diagnosed and 4 months after the WHO declared COVID-19 as a global pandemic, two pharma companies, Moderna and Pfizer, began phase III trials of mRNA COVID-19 vaccines. The FDA approval for use of the first mRNA vaccine (Pfizer's BNT162b2) only 5 months after phase II trials results showed high efficacy, and safety was met with both great enthusiasm and apprehension [36]. In light of reports of severe COVID-19 among parturients, the CDC added pregnancy to the list of high-risk conditions to prioritize vaccination, and the American College of Obstetricians and Gynecologists (ACOG) recommends not withholding vaccination from pregnant women at any stage of the pregnancy [37]. The enthusiasm surrounding the vaccine rollout was accompanied by unsubstantiated rumors, spread via social media, suggesting that the vaccine may lead to female sterility [38]. Several studies looked at the effects of the anti-SARS-CoV-2 mRNA vaccines on male and female fertility, IVF, and pregnancy outcome. A study by Gonzalez et al. compared sperm quality before and 70 days post anti-SARS-CoV-2 mRNA vaccine [BNT162b2 (Pfizer-BioNTech) and mRNA-1273 (Moderna)] among healthy volunteers. Results showed similar semen volume and sperm concentration and improved motility post vaccination [39]. Assessment of the effect of the BNT162b2 on ovarian reserve as assessed by serum AMH 3 months post vaccination showed no difference [40]. There were two studies that compared IVF treatment outcome in vaccinated and unexposed fertility patients [23, 41]. The comparison included oocyte yield, mature oocyte yield, estrogen and progesterone production, fertilization rate, blastocyst formation rate, embryo quality based on morphology staging, or euploid embryo rate which showed no difference between the two groups. One of the studies had also compared clinical and ongoing pregnancy rate as well as early pregnancy loss, again showing no measurable differences [41]. Another study examined the risk for early pregnancy loss among women vaccinated shortly before or during pregnancy and found the rate to be within the expected age-specific range [42]. The "V-safe study" had also reported that the rate of fetal congenital malformations among patients vaccinated with either the BNT162b2 or mRNA-1273 vaccines was within the expected range (2.2%) [43]. Interestingly, similarly to reports from COVID-19 recoverees, patients vaccinated with mRNA anti-SARS-CoV-2 vaccines had also reported on menstrual cycle changes. However, the mean change in menstrual cycle length was only for 1 day, and no changes were recorded in the length of menses [44].

Conclusion

COVID-19 is the first global epidemic to be managed with a coordinated worldwide effort to limit its spread and life toll. This action included the development and deployment of several effective vaccines in an unprecedented short time frame. The fast-tracking of the deployment of the vaccines led to an unparalleled social movement of anti-vaccine activism. Early reports on severe morbidity and mortality among pregnant women as well as of a potential vertical transmission led regulatory bodies to approve vaccination of women immediately prior and during pregnancy with the mRNA vaccines. Since, despite the emergence of new variants, the clinical safety and efficiency of these vaccines remains. Also, despite initial concerns, other than a transient decline in sperm quality in COVID-19 affected males, the reproductive system, IVF treatment outcomes, and early pregnancy do not seem susceptible to direct detrimental effects of the SARS-CoV-2 virus.

References

1. COVID-19 Excess Mortality Collaborators. Estimating excess mortality due to the COVID-19 pandemic: a systematic analysis of COVID-19-related mortality, 2020–2021. Lancet. 2022;399(10334):1513–36.
2. Brown TM. The COVID-19 pandemic in historical perspective: an AJPH Dossier. Am J Public Health. 2021;111(3):402–4.
3. Caramaschi S, et al. Histopathological findings and clinicopathologic correlation in COVID-19: a systematic review. Mod Pathol. 2021;34(9):1614–33.
4. Iqbal FM, et al. Characteristics and predictors of acute and chronic post-COVID syndrome: a systematic review and meta-analysis. eClinicalMedicine. 2021;36:100899.

5. Overton EE, Goffman D, Friedman AM. The epidemiology of COVID-19 in pregnancy. Clin Obstet Gynecol. 2022;65(1):110–22.
6. Drosten C, et al. Severe acute respiratory syndrome: identification of the etiological agent. Trends Mol Med. 2003;9(8):325–7.
7. Ludwig S, Zarbock A. Coronaviruses and SARS-CoV-2: a brief overview. Anesth Analg. 2020;131(1):93–6.
8. Tortorici MA, Veesler D. Structural insights into coronavirus entry. Adv Virus Res. 2019;105:93–116.
9. Fehr AR, Perlman S. Coronaviruses: an overview of their replication and pathogenesis. Methods Mol Biol. 2015;1282:1–23.
10. Edenfield RC, Easley CA. Implications of testicular ACE2 and the renin–angiotensin system for SARS-CoV-2 on testis function. Nat Rev Urol. 2022;19(2):116–27.
11. Yuki K, Fujiogi M, Koutsogiannaki S. COVID-19 pathophysiology: a review. Clin Immunol. 2020;215:108427.
12. Hoffmann M, et al. SARS-CoV-2 cell entry depends on ACE2 and TMPRSS2 and is blocked by a clinically proven protease inhibitor. Cell. 2020;181(2):271–280 e8.
13. Jackson CB, et al. Mechanisms of SARS-CoV-2 entry into cells. Nat Rev Mol Cell Biol. 2022;23(1):3–20.
14. Valdes G, et al. Distribution of angiotensin-(1–7) and ACE2 in human placentas of normal and pathological pregnancies. Placenta. 2006;27(2–3):200–7.
15. Pringle KG, et al. The expression and localization of the human placental prorenin/renin–angiotensin system throughout pregnancy: roles in trophoblast invasion and angiogenesis? Placenta. 2011;32(12):956–62.
16. Chen CP, et al. Functional characterization of the human placental fusogenic membrane protein syncytin 2. Biol Reprod. 2008;79(5):815–23.
17. Frendo JL, et al. Direct involvement of HERV-W Env glycoprotein in human trophoblast cell fusion and differentiation. Mol Cell Biol. 2003;23(10):3566–74.
18. Galang RR, et al. Risk factors for illness severity among pregnant women with confirmed severe acute respiratory syndrome coronavirus 2 infection-surveillance for emerging threats to mothers and babies network, 22 state, local, and territorial health departments, 29 March 2020–5 March 2021. Clin Infect Dis. 2021;73(Suppl 1):S17–23.
19. Kenigsberg S, Bentov Y. Does contemporary ART lead to pre-eclampsia? A cohort study and meta-analysis. J Assist Reprod Genet. 2021;38(3):651–9.
20. Blumenfeld Z. Possible impact of COVID-19 on fertility and assisted reproductive technologies. Fertil Steril. 2020;114(1):56–7.
21. Heiat M, et al. Phobia and fear of COVID-19: origins, complications and management, a narrative review. Ann Ig. 2021;33(4):360–70.
22. Herrero Y, et al. SARS-CoV-2 infection negatively affects ovarian function in ART patients. Biochim Biophys Acta Mol Basis Dis. 2022;1868(1):166295.
23. Bentov Y, et al. Ovarian follicular function is not altered by SARS-CoV-2 infection or BNT162b2 mRNA COVID-19 vaccination. Hum Reprod. 2021;36(9):2506–13.
24. Youngster M, et al. IVF under COVID-19: treatment outcomes of fresh ART cycles. Hum Reprod. 2022;37(5):947–53.
25. Shaw J, et al. Universal SARS-CoV-2 polymerase chain reaction screening and assisted reproductive technology in a coronavirus disease 2019 pandemic epicenter: screening and cycle outcomes from a New York City fertility center. Fertil Steril. 2021;116(4):980–7.
26. Carp-Veliscu A, et al. The effects of SARS-CoV-2 infection on female fertility: a review of the literature. Int J Environ Res Public Health. 2022;19(2):984.
27. Yang M, et al. Effects of SARS-CoV-2 infection on human reproduction. J Mol Cell Biol. 2021;13(10):695–704.
28. Delaroche L, et al. Evaluation of SARS-CoV-2 in semen, seminal plasma, and spermatozoa pellet of COVID-19 patients in the acute stage of infection. PLoS One. 2021;16(12):e0260187.
29. Best JC, et al. Evaluation of SARS-CoV-2 in human semen and effect on total sperm number: a prospective observational study. World J Mens Health. 2021;39(3):489–95.

30. Setti PEL, et al. First trimester pregnancy outcomes in a large IVF center from the Lombardy County (Italy) during the peak COVID-19 pandemic. Sci Rep. 2021;11(1):16529.
31. Musa SS, et al. Vertical transmission of SARS-CoV-2: a systematic review of systematic reviews. Viruses. 2021;13(9):1877.
32. Engels Calvo V, et al. Perinatal outcomes of pregnancies resulting from assisted reproduction technology in SARS-CoV-2-infected women: a prospective observational study. Fertil Steril. 2021;116(3):731–40.
33. Cavalcante MB, et al. COVID-19 and miscarriage: from immunopathological mechanisms to actual clinical evidence. J Reprod Immunol. 2021;148:103382.
34. Yang J, et al. COVID-19 pandemic and population-level pregnancy and neonatal outcomes in general population: a living systematic review and meta-analysis. Acta Obstet Gynecol Scand. 2022;101(3):273–92.
35. Peivandi S. Is genetic counseling necessary for pregnant women with coronavirus infection? An unresolved question. Taiwan J Obstet Gynecol. 2021;60(4):798.
36. Akamine CM, El Sahly HM. Messenger ribonucleic acid vaccines for severe acute respiratory syndrome coronavirus-2—a review. Transl Res. 2022;242:1–19.
37. ACOG Expert Work Group in Collaboration, Riley LE, Jamieson DJ, Hughes BL, Swamy G, Eckert LO, Turrentine M, Carroll S. Vaccinating pregnant and lactating patients against COVID-19. ACOG Obestet Gynecol. 2020.
38. Evans MB, et al. COVID-19 vaccine and infertility: baseless claims and unfounded social media panic. Washington, DC: American Society for Reproductive Medicine; 2021. Available at: https://www.fertstertdialog.com/posts/covid-19-vaccine-and-infertility-baseless-claims-and-unfounded-social-media-panic. Retrieved March 9, 2021.
39. Gonzalez DC, et al. Sperm parameters before and after COVID-19 mRNA vaccination. JAMA. 2021;326(3):273–4.
40. Mohr-Sasson A, et al. The effect of Covid-19 mRNA vaccine on serum anti-Mullerian hormone levels. Hum Reprod. 2022;37(3):534–41.
41. Aharon D, et al. In vitro fertilization and early pregnancy outcomes after coronavirus disease 2019 (COVID-19) vaccination. Obstet Gynecol. 2022;139(4):490–7.
42. Zauche LH, et al. Receipt of mRNA Covid-19 vaccines and risk of spontaneous abortion. N Engl J Med. 2021;385(16):1533–5.
43. Shimabukuro TT, et al. Preliminary findings of mRNA Covid-19 vaccine safety in pregnant persons. N Engl J Med. 2021;384(24):2273–82.
44. Edelman A, et al. Association between menstrual cycle length and coronavirus disease 2019 (COVID-19) vaccination: a U.S. Cohort. Obstet Gynecol. 2022;139(4):481–9.

COVID-19 Pandemic and Vaccination

2

Yonatan Oster

Introduction

The severe acute respiratory syndrome coronavirus 2 (SARS-CoV-2) virus was first described as the causative agent of acute respiratory infection in an outbreak in Wuhan, China, in the last months of 2019 [1]; this infection was later named coronavirus disease 2019 (COVID-19). The virus rapidly spread worldwide to create a global pandemic, including more than 500 million diagnosed infections and 6.3 million deaths, as of June 2022. The SARS-CoV-2 virus is a small RNA virus belongs to the coronaviruses family, which consists many viruses, mostly infecting animals but not humans. Other human affecting coronaviruses are very common and known to cause mild upper respiratory infections. Two of the viruses in this family have caused global outbreaks in the past, SARS-CoV-1 (previously SARS) in 2003 and the Middle Eastern respiratory syndrome coronavirus (MERS-CoV) which have caused several outbreaks since 2012. Contrary to most human coronaviruses who cause a very mild disease, the SARS-CoV-1 and MERS-CoV can cause severe disease with high mortality approaching 30%. The global spreading of COVID-19 has caused an immense effect on global health, finance, and geopolitics, and some of the data we know so far will be discussed in this chapter. Since its discovery in 2019, many mutations were found in the viral RNA, and accumulation of several mutations creates different viral variants that have distinct clinical and immunologic characteristics. The WHO has decided to classify and name the variants by Greek letters. The original virus is known as the wild type, and the common variants are the alpha, beta, delta, and omicron. The gradual emergence of these variants created the wavy pattern

Y. Oster (✉)
The Faculty of Medicine, Hebrew University of Jerusalem, Jerusalem, Israel

Department of Clinical Microbiology and Infectious Diseases, Hadassah Medical Center, Jerusalem, Israel
e-mail: yonatano@hadassah.org.il

J. G. Schenker et al. (eds.), *Hot Topics in Human Reproduction*, Reproductive Medicine for Clinicians 3, https://doi.org/10.1007/978-3-031-24903-7_2

of the pandemic, where a rise in morbidity is stopped by preventive measures and a certain immunity of the population, only to rise again after several months due to the development of a new variant [2]. In general, newer variants are more infective than their preceding and tend to cause a less severe disease.

Modes of Transmission

As a respiratory pathogen, the main mode of transmission of COVID-19 is through respiratory secretions. Similar to most respiratory pathogens, the virus particles can be found in respiratory droplets causing person-to-person transmission in close contact circumstances. The duration and distance during exposure to a positive person will affect the probability of transmission, so household exposure will more likely lead to infection than a short random outdoor encounter [3]. High viral loads are found in the first few days of infection, after 2–3 days of incubation, and levels decline after several days in immunocompetent persons. Additionally, two other modes of transmission were highly studied during the pandemic. First, as in the outbreaks caused by other coronaviruses, some virus particles can be spread through aerosol, which are much smaller droplets that can be transmitted to a longer distance and stay in the airborne for longer time [4]. Studies have shown that most of the virus is not carried by aerosol; however some medical procedures do, for example, endotracheal intubation and extubation, bronchoscopy, or open suctioning of airway secretions [5]. Another suspected mode of transmission is through contact, which is common in other respiratory infections, as virus-containing droplets fall on close surfaces and fomites, including hands, which can be also directly contaminated by touching the mouth or nose. This route is related to a small part of transmissions, mainly because the virus can stay infective only for several hours on dry surfaces [6]. The infectivity of the virus, measured by the average number of people any positive person infects (also called R0), is much higher than the Influenza virus or other common respiratory viruses [7]. A major reason for this high infectivity is the pre-symptomatic [8] and even asymptomatic transmission of COVID-19 [9], which assisted in the global spread of the pandemic.

Infection

SARS-CoV-2 is mainly a respiratory pathogen, as the virus invades and replicates in the mucosa of the naso- and oropharynx. In some cases, however, the infection progresses to the lower respiratory tract, to create pneumonia that can be mild but also severe and life-threatening [10, 11]. The symptoms of most patients are related to the upper and lower respiratory system and include cough, sore throat, and coryza, as well as systemic symptoms such as fever, malaise, headache, and loss of appetite. Loss of taste and/or smell was frequently reported with the first few variants but had become a rare phenomenon with the omicron variant. The virus can cause some manifestation out of the respiratory system, including myocarditis, gastrointestinal symptoms, and hyper-coagulation, mainly pulmonary emboli, stroke, and coronary disease. Risk factors of progression to severe disease are older age, immunosuppression (mainly in the humoral system), obesity, and chronic lung and cardiovascular pathologies. The main feature of severe COVID-19 is the acute respiratory distress syndrome (ARDS) complicated by multi-organ failure, mainly renal. Duration of symptoms in mild cases is usually 5 days or less in the immunocompetent, but can be much longer in immunocompromised patients. Some symptoms may persist longer, such as cough or loss of taste and/or smell. Post-COVID-19 symptoms, such as fatigue, shortness of breath, or memory loss, were described in many recovered patients. A rare inflammatory phenomenon, multisystem inflammatory syndrome, had developed in adults and children following COVID-19, resulting in high mortality [12].

Diagnosis

Clinical diagnosis of COVID-19 infection is challenging, as most patients present mild non-specific symptoms. Epidemiological data, such as rate of COVID-19 infection in a specific area, or close contact with a known case can assist in this diagnosis. The gold standard of laboratory testing is the real-time PCR, performed on nasopharyngeal and oropharyngeal swabs or on respiratory secretions. PCR tests have very high sensitivity and specificity, but are limited by high costs, long turnover times, and the need for expensive equipment and supplies. Lateral flow antigen tests provide rapid and cheap results and do not require trained laboratory technicians, but their sensitivity is 30–40% lower than for RT-PCR, depending on whether tested subjects are symptomatic [13]. This lower sensitivity has some advantages, as RT-PCR can detect residual viral RNA sometimes weeks after the infection. Serology tests are rarely used for diagnosis of acute infection, as IgM titers rise only 1–2 weeks following an infection. IgG antibody titers, however, can be useful in identifying previous infections. Anti-S antibodies can be found in previously infected and vaccinated persons, so anti-N antibodies should be measured to discriminate between these populations, as they do not rise after vaccination. Unfortunately, serologic tests cannot be used yet for assessment of protection from reinfections, as they represent only one part of a complex immune response. Additionally, antibodies against one variant might not be protecting against others.

Treatment

The majority of COVID-19 positive persons develop only a mild disease and require no therapy except bed rest and anti-pyretics. Sicker patients may need oxygen supplement and the use of noninvasive or invasive ventilation. These patients should get adequate supportive care as well. Due to the hyper-coagulation produced by the infection, all hospitalized patients are given prophylactic anticoagulation, mainly low-molecular-weight heparin, to prevent severe thrombotic complications [14].

Since the first appearance of SARS-CoV-2 in 2019, many anti-inflammatory medications were suggested to mitigate the disease progression. The most successful drug group is corticosteroids, and most of the moderate and severe patients receive them, mainly dexamethasone, which was proven in the multi-center, international, RECOVERY study [15]. Other options, such as anti-IL-6 agents, may have a role in some cases [16]. Secondary infections are described in many COVID-19 patients. Few are respiratory co-infections, which can be bacterial, viral, or fungal, others hospital-related infections [17].

Antivirals

The search for effective antiviral medications to treat COVID-19 is still in progress; however several such drugs have already been used on millions of patients. Remdesivir is an adenosine-analogue that stops viral RNA transcription. Its effectiveness was proven in shortening time to recovery in moderate and severe disease and in preventing disease progression in high-risk patients with mild disease [18]. It is administered intravenously, making its use in non-hospitalized patients difficult. Two oral antiviral agents, Paxlovid (ritonavir-boosted nirmatrelvir) and Lagevrio (Molnupiravir), are effective in preventing disease progression in high-risk patients, when given in the first 5 days of the disease [19, 20]. Several other agents were suggested as treatments; some existing medications such as ivermectin, hydroxychloroquine, or azithromycin; and some novel agents, which all failed in controlled clinical trials. Antibody transfusion, by either plasma of recovered patients [21, 22] or synthetic monoclonal antibodies [23], has been used in different clinical scenarios with mixed results, depending also on the current SARS-CoV-2 variant.

Prevention

Governments and health organizations worldwide have made many efforts to delay the spread of COVID-19 since its first appearance in China. These efforts include restriction on international travel, indoor and outdoor gatherings, and even work places, public transportation and schools, and even curfews in some countries. Most importantly, mandatory isolation of COVID-19-positive persons and quarantine of susceptible contacts were deployed in order to interrupt the chains of infection, including active tracing of contacts [24, 25]. Additionally, the use of facemasks by the general public was promoted by many authorities. Despite these and other measures, the virus did spread worldwide in the first months of 2020; however this global effort did slow down the progression of the pandemic and allowed the health services to function amidst the worst pandemic waves. Since some virus can be transferred via aerosol, some authorities have recommended the use of N95 or similar highly effective respirators; the difficulty in wearing these correctly and consistently and the demand for these by medical organizations support the use of regular (surgical) facemasks. Reusable cloth facemasks, while being better for the environment, were proven to be less effective [26]. Protective measures include contact and droplet protection, i.e., gown, gloves, facemask, and eye protection, with addition of N95 or similar respirators when exposure to aerosol is expected. The use of such equipment had increased the burden on the healthcare workers, but had saved many infections worldwide [27].

Vaccination

Immediately after the publication of the SARS-CoV-2 RNA sequence, the race for vaccine development began. By the end of 2020, several vaccines were available for use or at late production stages. Two vaccines, Comirnaty (BNT162b2) and Spikevax (mRNA-1273), were developed based on a novel mRNA vaccination technology and have shown very high effectiveness in preventing symptomatic disease, complications, and mortality in the first months after the first two doses [28–30]. Other vaccinations, such as the ChAdOx1, Ad26.COV2, Gam-COVID-Vac (known as Sputnik V), and others, are based on older and more studied vaccine technologies and have shown moderate-to-high effectiveness. Vaccine effectiveness and safety was proved primarily in adults, and the data had gradually expanded to younger ages. Special populations, including immunocompromised and pregnant persons, can also be vaccinated safely [31]. This protection starts to wane after several months, and a booster dose is necessary to maintain population immunity [32, 33]. Mass vaccination operations were performed in many countries, and billions of vaccine doses were given. Despite the high efficacy of the vaccines in preventing mortality and complications, their ability to prevent infection is limited, at best. Studies show that vaccinated individuals that were infected can secrete similar amounts of infective virus particles as the non-vaccinated. Additionally, vaccine efficacy is lower for newer variants, similar to the natural immunity of recovered persons. Newer and broader vaccinations are being developed. Common vaccine side effects include local pain, malaise, and even fever in mRNA vaccine recipients and regional lymph node enlargement or myocarditis in rare cases [34]. Vaccine-induced thrombotic thrombocytopenia is another rare syndrome, associated with the Ad26.COV2 and ChAdOx1 vaccines, characterized by venous or arterial thrombosis associated with thrombocytopenia and detectable anti-platelet factor 4 antibodies [35]. Another type of vaccines is passive immunizations. Evusheld, a mixture of two synthetic antibodies (Tixagevimab and Cilgavimab), can create effective protection in immunocompromised persons if given prior to their exposure to SARS-CoV-2. New vaccinations, with the ability to generate sustained protection against infection, are needed in order to eliminate this infection. It is more likely, however, that the SARS-CoV-2 will become a part of the ensemble of human respiratory infections.

References

1. Zhu N, Zhang D, Wang W, et al. A novel coronavirus from patients with pneumonia in China, 2019. N Engl J Med. 2020;382(8):727–33. https://doi.org/10.1056/NEJMoa2001017.
2. Chen Y, Liu Q, Zhou L, Zhou Y, Yan H, Lan K. Emerging SARS-COV-2 variants: why, how, and what's next? Cell Insight. 2022;1(3):100029. https://doi.org/10.1016/j.cellin.2022.100029.
3. Oster Y, Benenson S, Yochi Harpaz L, et al. Association between exposure characteristics and the risk for COVID-19 infection among health care workers with and without BNT162b2 vaccination. JAMA Netw Open. 2021;4(9):e2125394. https://doi.org/10.1001/jamanetworkopen.2021.25394.
4. Edwards DA, Ausiello D, Salzman J, et al. Exhaled aerosol increases with COVID-19 infection, age, and obesity. Proc Natl Acad Sci U S A. 2021;118(8):830118. https://doi.org/10.1073/pnas.2021830118.
5. Klompas M, Baker M, Rhee C. What Is an Aerosol-Generating Procedure? JAMA Surg. 2021;156(2):113–14. https://doi.org/10.1001/jamasurg.2020.6643.
6. Meyerowitz EA, Richterman A, Gandhi RT, Sax PE. Transmission of SARS-CoV-2: A Review of Viral, Host, and Environmental Factors. Ann Intern Med. 2021;174(1):69–79. https://doi.org/10.7326/M20-5008.
7. Liu Y, Gayle AA, Wilder-Smith A, Rocklöv J. The reproductive number of COVID-19 is higher compared to SARS coronavirus. J Travel Med. 2020;27(2):taaa021. https://doi.org/10.1093/jtm/taaa021.
8. Arons MM, Hatfield KM, Reddy SC, et al. Presymptomatic SARS-CoV-2 infections and transmission in a skilled nursing facility. N Engl J Med. 2020;382(22):2081–90. https://doi.org/10.1056/NEJMoa2008457.
9. Bai Y, Yao L, Wei T, et al. Presumed asymptomatic carrier transmission of COVID-19. JAMA. 2020;323(14):1406–7. https://doi.org/10.1001/jama.2020.2565.
10. Guan WJ, Ni ZY, Hu Y, et al. Clinical characteristics of coronavirus disease 2019 in China. N Engl J Med. 2020;382(18):1708–20. https://doi.org/10.1056/NEJMoa2002032.
11. Wu Z, McGoogan JM. Characteristics of and important lessons from the coronavirus disease 2019 (COVID-19) outbreak in China: summary of a report of 72,314 cases from the Chinese Center for Disease Control and Prevention. JAMA. 2020;323(13):1239–42. https://doi.org/10.1001/jama.2020.2648.
12. Feldstein LR, Rose EB, Horwitz SM, et al. Multisystem inflammatory syndrome in U.S. children and adolescents. N Engl J Med. 2020;383(4):334–46. https://doi.org/10.1056/NEJMoa2021680.
13. Dinnes J, Deeks JJ, Berhane S, et al. Rapid, point-of-care antigen and molecular-based tests for diagnosis of SARS-CoV-2 infection. Cochrane Database Syst Rev. 2021;3(3):CD013705. https://doi.org/10.1002/14651858.CD013705.pub2.
14. Flumignan RLG, Civile VT, Tinôco JD, et al. Anticoagulants for people hospitalised with COVID-19. Cochrane Database Syst Rev. 2022;3:CD013739. https://doi.org/10.1002/14651858.CD013739.pub2.
15. RECOVERY Collaborative Group, Horby P, Lim WS, et al. Dexamethasone in hospitalized patients with Covid-19. N Engl J Med. 2021;384(8):693–704. https://doi.org/10.1056/NEJMoa2021436.
16. Lescure FX, Honda H, Fowler RA, et al. Sarilumab in patients admitted to hospital with severe or critical COVID-19: a randomised, double-blind, placebo-controlled, phase 3 trial. Lancet Respir Med. 2021;9(5):522–32. https://doi.org/10.1016/S2213-2600(21)00099-0.
17. Lansbury L, Lim B, Baskaran V, Lim WS. Co-infections in people with COVID-19: a systematic review and meta-analysis. J Infect. 2020;81(2):266–75. https://doi.org/10.1016/j.jinf.2020.05.046.
18. Ansems K, Grundeis F, Dahms K, et al. Remdesivir for the treatment of COVID-19. Cochrane Database Syst Rev. 2021;8:CD014962. https://doi.org/10.1002/14651858.CD014962.
19. Hammond J, Leister-Tebbe H, Gardner A, et al. Oral nirmatrelvir for high-risk, nonhospitalized adults with Covid-19. N Engl J Med. 2022;386(15):1397–408. https://doi.org/10.1056/NEJMoa2118542.

20. Jayk Bernal A, Gomes da Silva MM, Musungaie DB, et al. Molnupiravir for oral treatment of Covid-19 in nonhospitalized patients. N Engl J Med. 2022;386(6):509–20. https://doi.org/10.1056/NEJMoa2116044.
21. Chen L, Xiong J, Bao L, Shi Y. Convalescent plasma as a potential therapy for COVID-19. Lancet Infect Dis. 2020;20(4):398–400. https://doi.org/10.1016/S1473-3099(20)30141-9.
22. Piechotta V, Iannizzi C, Chai KL, et al. Convalescent plasma or hyperimmune immunoglobulin for people with COVID-19: a living systematic review. Cochrane Database Syst Rev. 2021;5:CD013600. https://doi.org/10.1002/14651858.CD013600.pub4.
23. Hirsch C, Park YS, Piechotta V, et al. SARS-CoV-2-neutralising monoclonal antibodies to prevent COVID-19. Cochrane Database Syst Rev. 2022;6:CD014945. https://doi.org/10.1002/14651858.CD014945.pub2.
24. Hellewell J, Abbott S, Gimma A, et al. Feasibility of controlling COVID-19 outbreaks by isolation of cases and contacts [published correction appears in Lancet Glob Health. 2020 Mar 5]. Lancet Glob Health. 2020;8(4):e488–96. https://doi.org/10.1016/S2214-109X(20)30074-7.
25. Kucharski AJ, Klepac P, Conlan AJK, et al. Effectiveness of isolation, testing, contact tracing, and physical distancing on reducing transmission of SARS-CoV-2 in different settings: a mathematical modelling study. Lancet Infect Dis. 2020;20(10):1151–60. https://doi.org/10.1016/S1473-3099(20)30457-6.
26. Andrejko KL, Pry JM, Myers JF, et al. Effectiveness of Face Mask or Respirator Use in Indoor Public Settings for Prevention of SARS-CoV-2 Infection - California, February-December 2021. MMWR Morb Mortal Wkly Rep. 2022;71(6):212–16. Published 2022 Feb 11. https://doi.org/10.15585/mmwr.mm7106e1.
27. Su CY, Peng CY, Liu HL, Yeh IJ, Lee CW. Comparison of Effects of N95 Respirators and Surgical Masks to Physiological and Psychological Health among Healthcare Workers: A Randomized Controlled Trial. Int J Environ Res Public Health. 2021;18(24):13308. Published 2021 Dec 17. https://doi.org/10.3390/ijerph182413308.
28. Lindsey R, Baden Hana M, El Sahly Brandon, et al. Efficacy and Safety of the mRNA-1273 SARS-CoV-2 Vaccine. N Engl J Med. 2021;384(5):403–16. https://doi.org/10.1056/NEJMoa2035389.
29. Polack FP, Thomas SJ, Kitchin N, et al. Safety and efficacy of the BNT162b2 mRNA Covid-19 vaccine. N Engl J Med. 2020;383(27):2603–15. https://doi.org/10.1056/NEJMoa2034577.
30. Thomas SJ, Moreira ED Jr, Kitchin N, et al. Safety and efficacy of the BNT162b2 mRNA Covid-19 vaccine through 6 months. N Engl J Med. 2021;385(19):1761–73. https://doi.org/10.1056/NEJMoa2110345.
31. Wainstock T, Yoles I, Sergienko R, Sheiner E. Prenatal maternal COVID-19 vaccination and pregnancy outcomes. Vaccine. 2021;39(41):6037–40. https://doi.org/10.1016/j.vaccine.2021.09.012.
32. Ferdinands JM, Rao S, Dixon BE, et al. Waning 2-dose and 3-dose effectiveness of mRNA vaccines against COVID-19-associated emergency department and urgent care encounters and hospitalizations among adults during periods of delta and omicron variant predominance—VISION network, 10 states, August 2021–January 2022. MMWR Morb Mortal Wkly Rep. 2022;71(7):255–63. https://doi.org/10.15585/mmwr.mm7107e2.
33. Regev-Yochay G, Gonen T, Gilboa M, et al. Efficacy of a fourth dose of Covid-19 mRNA vaccine against omicron. N Engl J Med. 2022;386(14):1377–80. https://doi.org/10.1056/NEJMc2202542.
34. Mevorach D, Anis E, Cedar N, et al. Myocarditis after BNT162b2 vaccination in Israeli adolescents. N Engl J Med. 2022;386(10):998–9. https://doi.org/10.1056/NEJMc2116999.
35. Klok FA, Pai M, Huisman MV, Makris M. Vaccine-induced immune thrombotic thrombocytopenia. Lancet Haematol. 2022;9(1):e73–80. https://doi.org/10.1016/S2352-3026(21)00306-9.

Pregnancy: Ethical Issues of Vaccine Refusal

3

Avraham Steinberg

Good ethics start with good facts. To discuss the ethical dilemmas concerning COVID-19 vaccination during pregnancy, it is first required to analyze the medical-scientific available data regarding the outcome of COVID-19 infection during pregnancy upon the pregnant individual and upon the fetus and the efficacy and safety of COVID-19 vaccines upon both the pregnant individual and the fetus.

Medical-Scientific Aspects

COVID-19 Pandemic and Vaccination

The COVID-19 pandemic started in China in December 2019 and rapidly spread to over 200 countries worldwide. As of the beginning of March 2022, 446 billion people were infected, and over 6 billion people lost their lives due to the pandemic.

All preventative measures failed to stop the pandemic. Soon after its outbreak, it was clear that vaccinating the entire world population will be the most effective tool to fight the pandemic. As of the beginning of March 2022, over ten billion doses of COVID-19 vaccines have been administered worldwide. This constitutes over 62% of the world population.

Several COVID-19 vaccines, including the Pfizer-BioNTech and Moderna vaccines which are based on mRNA—a relatively new model of vaccines—have been successfully developed.

On 2 December 2020, the United Kingdom's regulatory agency (MHRA) gave temporary regulatory approval for the Pfizer-BioNTech vaccine, becoming the first country to approve the vaccine [1, 2]. On 11 December 2020, the FDA granted an

A. Steinberg (✉)
Medical Ethics Unit, Shaare Zedek Medical Center, Jerusalem, Israel
e-mail: steinberg@e-tal.org

J. G. Schenker et al. (eds.), *Hot Topics in Human Reproduction*, Reproductive Medicine for Clinicians 3, https://doi.org/10.1007/978-3-031-24903-7_3

emergency use authorization (EUA) for the Pfizer-BioNTech COVID-19 vaccine. A week later, they granted an EUA for mRNA-1273, the Moderna vaccine [3].

Currently, 30 vaccines are authorized by at least 1 national regulatory authority for public use. Over 200 vaccines are undergoing clinical trials that have yet to be authorized.

COVID-19 Illness and Vaccination in Pregnancy

Globally, over 200 million individuals are pregnant each year. This creates a very significant and special group in relation to the pandemic and to vaccination.

Several studies demonstrated the risks of morbidity and mortality associated with COVID-19 in pregnancy on maternal and neonatal outcomes compared with not-infected pregnant individuals.

In a multinational cohort study, COVID-19 in pregnancy was associated with consistent and substantial increases in severe maternal morbidity and mortality, as well as neonatal complications when pregnant individuals with and without COVID-19 diagnosis were compared. Pregnant individuals with COVID-19 diagnosis were at higher risk for preeclampsia/eclampsia, severe infections, intensive care unit admission, maternal mortality, preterm birth, medically indicated preterm birth, severe neonatal morbidity index, and severe perinatal morbidity and mortality index. Asymptomatic pregnant individuals with COVID-19 diagnosis remained at a higher risk only for maternal morbidity and preeclampsia [4]. Several other studies corroborated these results [5–8]. In the United States, as of the end of February 2022, there were 182,847 infected pregnant individuals, 29,519 were hospitalized, and 285 died. Nearly all the COVID-19 cases among pregnant individuals to date have been among unvaccinated persons [9].

Studies have also documented serious consequences for the fetus and the newborn. These include stillbirth and early delivery due to maternal decompensation, leading to complicated neonatal courses and multiple neonatal deaths [6, 8].

Although pregnant individuals and their fetuses are at a higher risk of severe complications of COVID-19 infection, clinical trials for the available vaccines excluded pregnant and lactating women. Pregnant individuals were excluded from the initial phase 3 clinical trials of COVID-19 vaccines by Pfizer and Moderna.

Exclusion of pregnant individuals from clinical trials testing the safety and efficacy of COVID-19 vaccines is occurring even though, over the past two decades, several advisory bodies and ethics experts have issued recommendations for including pregnant individuals in clinical trials [10]. Hence, widespread failure to appropriately include pregnant women in vaccine research means that evidence about safety and efficacy in pregnancy has been limited and late in coming. As a result, in many countries, pregnant individuals have been denied the opportunity to receive COVID-19 vaccines that would have protected them and their offspring from the ravages of this disease.

Nonetheless, the consequences of the COVID-19 disease in pregnancy prompted many healthcare organizations and countries to support vaccination in pregnancy.

In Israel, the Ministry of Health and Vaccines Prioritization Committee recommended vaccination for pregnant individuals. Some jurisdictions in the United States are already offering the vaccine to pregnant individuals, including the District of Columbia, Pennsylvania, and Mississippi.

Professional societies, such as the Centers for Disease Control and Prevention (CDC), the American College of Obstetricians and Gynecologists (ACOG), the Society for Maternal-Fetal Medicine (SMFM), and the Royal College of Obstetricians and Gynaecologists (RCOG), all support COVID-19 vaccination in pregnancy since the benefits outweigh the risks [11, 12].

Based on the current available clinical data, vaccination during pregnancy has been proved to be effective and safe, both for the pregnant individual and for the fetus; the benefits of vaccination during pregnancy outweigh potential risks [13, 14].

Several studies discussed the attitude toward and knowledge of pregnant and nonpregnant individuals regarding the COVID-19 disease and vaccination and their effect on pregnant individuals, the fetus, and the newborn. These studies evaluated the sociodemographic characteristics, vaccination history, perception of risk for the COVID-19 pandemic, the impact of the COVID-19 pandemic, and acceptance or refusal of the COVID-19 vaccination during pregnancy.

The percentage of vaccine acceptance among pregnant women varied greatly among the surveys, ranging from 16 to 52% [15–18].

Most common refusal reasons were lack of data about COVID-19 vaccine safety in pregnant populations and the possibility of harm to the fetus. Pregnant individuals in the first trimester expressed higher acceptance of COVID-19 vaccination than those in the second and third trimesters [15]. Receipt of influenza vaccine during the previous season was associated with higher odds of vaccine acceptability [16].

Social-Ethical Aspects

Modern medical bioethics operates according to four fundamental principles: autonomy, beneficence, non-maleficence, and justice [19].

Beneficence is defined as the obligation of a physician to act for the benefit of the patient. The principle calls for not just avoiding harm but also to actively benefit patients and to promote their welfare.

Non-maleficence is defined as the obligation of a physician not to harm the patient.

The philosophical underpinning for autonomy is that all persons have intrinsic and unconditional worth and, therefore, should have the power to make rational decisions and moral choices, and each person should be allowed to exercise his or her wishes for self-determination.

The principle of autonomy became an overriding principle, and it replaced almost totally the long-standing patient–physician relationship based on paternalism. Nonetheless, as is true for all four principles, autonomy needs to be weighed against competing moral principles and in some instances may be overridden.

The evidence-based professional data presented in part Ib provides unequivocal and clear guidance: Physicians should recommend COVID-19 vaccination to persons who are pregnant.

The grave consequences resulting from COVID-19 infection during pregnancy, and the efficacy and safety of the vaccinations, constitute a prima facia moral obligation upon healthcare providers to advise, promote, and recommend vaccination at all stages of pregnancy. This is based on the moral principles of beneficence and non-maleficence.

Safety information on COVID-19 vaccines must be clearly communicated to pregnant individuals to provide reassurance and facilitate informed pregnancy vaccine decisions. Targeted interventions to promote COVID-19 vaccine uptake among ethnic minority and lower-income individuals may be needed [20]. Medical information about the safety, effectiveness, and benefits of vaccinations among pregnant individuals was found to increase the acceptance of vaccinations by 105.6% Polish patients and by 176% among Ukraine patients [18].

The strongest predictors of vaccine acceptance included confidence in vaccine safety or effectiveness, worrying about COVID-19 disease, belief in the importance of vaccines to their own country, compliance to mask guidelines, trust of public health agencies/health science, as well as attitudes toward routine vaccines [17].

Various terms are used to define the fact that certain people do not agree to be vaccinated: objection, hesitation, and refusal. Objection refers to a conscientious opposition to vaccines at large or to a particular vaccine; hesitancy refers to an attitude such as doubts or concerns [17]. Refusal refers to the end result of either objection or hesitancy.

Refusal and opposition to vaccines is not a new phenomenon. It appeared soon after the introduction of the first vaccine—the smallpox vaccine—in the eighteenth century.

The beliefs and arguments of the anti-vaccine movements have remained unchanged in the past two centuries, but new social media has facilitated the dissemination of information against vaccines [21].

There is a moral responsibility upon the pregnant individual to protect the fetus and her own health and well-being. Hence, the individual should accept the legitimate recommendation to be vaccinated.

Yet, based on the principle of autonomy, society ought to respect the right of an individual to make rational decisions and moral choices, and each person should be allowed to exercise his or her wishes for self-determination. Therefore, a pregnant individual who refuses to be vaccinated should not be coerced to do so.

Conclusions

COVID-19 can have severe consequences in pregnancy: evidence indicates that pregnant individuals with COVID-19 are at increased risk of severe illness and death, of giving birth to preterm babies, and potentially of other adverse pregnancy outcomes such as stillbirth.

COVID-19 vaccines are highly effective: They provide strong protection against severe illness and deaths. Pregnant individuals are likely to receive the same level of protection from the vaccines as nonpregnant people.

Safety data in pregnancy are increasing and are reassuring: Evidence on the safety of COVID-19 vaccines during pregnancy has been growing. To date, animal studies, monitoring of pregnant individuals who have received the vaccines, and experience using vaccines with similar components have not identified any pregnancy-specific safety concerns [14].

Based on the principles of beneficence and non-maleficence, it is the moral obligation upon healthcare providers to advocate, promote, and recommend vaccination for pregnant individuals.

Based on the principle of autonomy, a pregnant individual who refuses to be vaccinated should not be coerced to do so.

References

1. Medicines and Healthcare Products Regulatory Agency (MHRA). Information for Healthcare Professionals on Pfizer/BioNTech COVID-19 vaccine. Stepney: Medicines and Healthcare Products Regulatory Agency (MHRA); 2020.
2. Medicines and Healthcare Products Regulatory Agency (MHRA). UK medicines regulator gives approval for first UK COVID-19 vaccine. Stepney: Medicines and Healthcare Products Regulatory Agency, Government of the UK; 2020.
3. Oliver SE, Gargano JW, Marin M, Wallace M, Curran KG, Chamberland M, et al. The advisory committee on immunization practices' interim recommendation for use of Moderna COVID-19 vaccine—United States, December 2020. MMWR. 2021;69(5152):1653–6. https://doi.org/10.15585/mmwr.mm695152e1.
4. Villar J, Ariff S, Gunier RB, Thiruvengadam R, Rauch S, Kholin A, Roggero P, et al. Maternal and neonatal morbidity and mortality among pregnant women with and without COVID-19 infection. The INTERCOVID multinational cohort study. JAMA Pediatr. 2021;175(8):817–26. https://doi.org/10.1001/jamapediatrics.2021.1050.
5. Ko JY, DeSisto CL, Simoeon RM, Ellington S, Galang RR, et al. Adverse pregnancy outcomes, maternal complications, and severe illness among US delivery hospitalization with and without coronavirus disease 2019 (COVID-19) diagnosis. Clin Infect Dis. 2021;73(Supplement_1):S24–31. https://doi.org/10.1093/cid/ciab344.
6. Allotey J, Stallings E, Bonet M, et al. Clinical manifestations, risk factors, and maternal and perinatal outcomes of coronavirus disease 2019 in pregnancy: living systematic review and meta-analysis. BMJ. 2020;370:m3320. https://doi.org/10.1136/bmj.m3320.
7. Wei SQ, Bilodeau-Bertrand M, Liu S, et al. The impact of COVID-19 on pregnancy outcomes: a systematic review and meta-analysis. CMAJ. 2021;193(16):E540–E48. https://doi.org/10.1503/cmaj.202604.
8. Chmielewska B, Barratt I, Townsend R, Kalafat E, van der Meulen J, Gurol-Urganci I, O'Brien P, et. al. Effects of the COVID-19 pandemic on maternal and perinatal outcomes: a systematic review and meta-analysis. Lancet Glob Health. 2021;9:e759–72.
9. CDC Data Tracker, Pregnancy Data, 2021.
10. Farrell R, Michie M, Pope R. Pregnant women in trials of COVID-19: a critical time to consider ethical frameworks of inclusion in clinical trials. Ethics Hum Res. 2010;42(4):17–23. https://doi.org/10.1002/eahr.500060.
11. Rasmussen SA, Kelley CF, Horton JP, Jamieson DJ. Coronavirus disease 2019 (COVID-19) vaccines and pregnancy: what obstetricians need to know. Obstet Gynecol. 2021;137(3):408–14. https://doi.org/10.1097/AOG.0000000000004290.

12. Stafford IA, Parchem JG, Sibai BM. The coronavirus disease 2019 vaccine in pregnancy: risks, benefits, and recommendations. Am J Obstet Gynecol. 2021;224(5):484–95. https://doi.org/10.1016/j.ajog.2021.01.022.
13. Burd I, Kino T, Segars J. The Israeli study of Pfizer BNT162b2 vaccine in pregnancy: considering maternal and neonatal benefits. J Clin Invest. 2021;131(13):e150790. https://doi.org/10.1172/JCI150790.
14. World Health Organization. Questions and answers: COVID-19 vaccines and pregnancy. Geneva: World Health Organization; 2022.
15. Ayhan SG, Oluklu D, Atalay A, Beser DM, Tanacan A, Tekin OM, Sahin D. COVID-19 vaccine acceptance in pregnant women. Int J Gynaecol Obstet. 2021;154(2):291–6. https://doi.org/10.1002/ijgo.13713.
16. Battarbee AM, Stockwell MS, Varner M, Newes-Adeyi G, Daugherty M, Gyamfi-Bannerman C, Tita AT, et al. Attitudes toward COVID-19 illness and COVID-19 vaccination among pregnant women: a cross-sectional multicenter study during August–December 2020. Am J Perinatol. 2022;39(1):75–83. https://doi.org/10.1055/s-0041-1735878.
17. Skjefte M, Ngirbabul M, Akeju O, Escudero D, Hernandez-Diaz S, Wyszynski DF, Wu JW. COVID-19 vaccine acceptance among pregnant women and mothers of young children: results of a survey in 16 countries. Eur J Epidemiol. 2021;36(2):197–211. https://doi.org/10.1007/s10654-021-00728-6.
18. Januszek S, Siwiec N, Januszek R, Kluz M, Lebed R, Toś P, Góra T, Plens K, et al. Approach of pregnant women from Poland and the Ukraine to COVID-19 vaccination-the role of medical consultation. Vaccines (Basel). 2022;10(2):255. https://doi.org/10.3390/vaccines10020255.
19. Beauchamp TL, Childress JF, editors. Principles of biomedical ethics. 8th ed. New York: Oxford University Press; 2019.
20. Skirrow H, Barnett S, Bell S, Riaposova L, Mounier-Jack S, Kampmann B, Holder B. Women's views on accepting COVID-19 vaccination during and after pregnancy, and for their babies: a multi-methods study in the UK. BMC Pregnancy Childbirth. 2022;22(1):33. https://doi.org/10.1186/s12884-021-04321-3.
21. de Menezes Succi RC. Vaccine refusal—what we need to know. J Pediatr. 2018;94(6):574–81. https://doi.org/10.1016/j.jped.2018.01.008.

Part II

Beginning of Human Life

4 Scientific and Religious Controversies on Beginning of Human Life

Asim Kurjak

One of the most controversial topics in modern bioethics, science, and philosophy is the beginning of individual human life. In the seemingly endless debate, strongly stimulated by recent technologic advances in human reproduction, a synthesis between scientific data and hypothesis, philosophical thought, and issues of humanities has become a necessity to deal with ethical, juridical, and social problems. Furthermore, in this field there is a temptation to ask science to choose between opinions and beliefs, which neutralize one another. The question of when human life begins requires the essential aid of different forms of knowledge. Here we become involved in the juncture between science and religion, which needs to be carefully explored.

Modern bioethics and science are strongly concerned for the respect of human life at both ends of its existence (birth and death), but other sciences (e.g., philosophy, technology, psychology, sociology, law, and politics) consider the beginning of human life according to different points of view. However, bioethical topics like this one cannot be treated from only one perspective (e.g., biological, philosophical, or religious) because conclusions might be not good enough or reductive. This reality should be regarded in all its richness: an embryo gives a biologist and a geneticist substance for consideration, but because we are talking about the beginning of human life, it requires philosophical–anthropological consideration and confrontation with theology; in its protection we have to include ethics and law. In experiencing and investigating social behavior, other disciplines, such as the history of medicine and sociology, have to be included.

A. Kurjak (✉)
Obstetrics and Gynecology at Medical School Universities of Zagreb and Sarajevo, Zagreb, Croatia

Sarajevo School of Science and Technology, Sarajevo, Bosnia and Herzegovina

J. G. Schenker et al. (eds.), *Hot Topics in Human Reproduction*, Reproductive Medicine for Clinicians 3, https://doi.org/10.1007/978-3-031-24903-7_4

It is hard to answer the question when human life should be legally protected. At the time of conception? At the time of implantation? At the time of birth? In all countries (except Ireland and Liechtenstein), juridical considerations are based on Roman law. Roman civil law says that the fetus has right when it is born or if it is born-nasciturus.

Few countries agree with definition of beginning of human personality at the time of conception. The majority does not grant legal status to the human embryo in vitro (i.e., during the 14 days after fertilization). Thus, even in the absence of legal rights, there is no denying that the embryo constitutes the beginning of human life, a member of the human family. Therefore, whatever the attitude, every country has to examine which practices are compatible with the respect of that dignity and the security of human genetic material.

The question when a human life begins and how to define it could be answered only through the inner-connecting pathways of history, philosophy, medical science, and religion. It has not been easy to determine where to draw the fine line between the competence of science and metaphysics in this delicate philosophical field. To a large extent, the drawing of this line depends on one's fundamental philosophical outlook. The point at which human life begins will always be seen differently by different individuals, groups, cultures, and religious faiths. In democracy there are always at least two sides, and the center holds only when the majority realizes that without a minority democracy itself is lost. The minority in turn must realize its best chance lies in persuasion by reason and thoughtfulness rather than fanaticism.

In recent years, we have noticed robust increase of interest in the relationship between science and religion worldwide. In the past, the abovementioned, more or less autonomous intellectual activities often tried to dominate one over another, or they ignored each other. Only in recent times, most scientists and many theologians accept the view that scientific and religious "truths" are complementary and thus only methodologically independent. Today, science and religion are an important factor in the life of the people, the country, and the world. Science along with religion is the greatest gift the almighty granted on us.

Anticipating the future relations between science and technology, we can only extrapolate and wonder. In this century, we human beings have come to know who we are and where we are in ways unprecedented in all past millennia. We know the size, age, and extent of our universe; we know the deep evolutionary history of our planet and ourselves as part of this story. These facts of science have required integration into our classical religious worldviews; and these blending of theory and principle in science and religion will continue. In this century, we human beings have gained, through science and technology, more power than ever before to affect, for better or worse, our own well-being, that of the human and natural worlds, and even planetary history. The fate of the Earth, the fate of all who dwell thereon, depends, in the next century, on the responsible use of that power. Everything depends on how we join science, ethics, and religion in practice [1].

Modern science is not interested in the nature, but what we can say about nature; one does not invent occurrences yet it interprets it. Science differs from religion in a way that its truth can and must be experimentally verified and its methodological knowledge can be learned. Religion is dominated by irrational moment and science by rational moment. Intellectual knowledge in science is expressed quantitatively, in the form of mathematical formulas and equations, but in religion qualitatively in the form of metaphors/abstractions. Technologies as practical expression of science on the other hand worship as practical expression of theology, and society of laymen, as the basis of democracy, today represent the pragmatic Western system of real capitalism. Today more often than ever, a dramatic development of technology opens a range of possibilities, which are all generated by science, but often aren't sufficiently analyzed and all the alternatives and consequences aren't understood. Most of the alternatives and consequences are not even possible to be observed and understood, and they remain exclusively in the domain of science and technology. Our life is far richer than it can be described only by science, and therefore the views of the great religions certainly must be taken in consideration.

At present, for instance, there is more dialogue and integration in physics, ample conflict, and considerable independence between biology and religion. Whether that trend will continue depends partly on discoveries as yet unknown in physics, astronomy, and molecular and evolutionary biology.

Global leadership in science and technology has not translated into leadership in infant health, life expectancy, rates of literacy, equality of opportunity, productivity of workers, or efficiency of resource consumption. Neither has it overcome failing education systems, decaying cities, environmental degradation, unaffordable health care, and the largest national debt in history. Basic human needs—elemental needs—are intrinsically different from other material needs because they can be satisfied. Other needs appear to be insatiable, as the consumption patterns of the United States clearly demonstrate. Once basic human needs are met, satisfaction with our lives cannot be said to depend on the amount of things we acquire, use, and consume. More technology-based economic growth is not necessary to satisfy humanity's elemental needs, nor does more growth quench our thirst for consumption. In terms of the social contract, we justify more growth because it is supposedly the most efficient way to spread economic opportunity and social well-being. I am suggesting that this reasoning is simplistic and often specious [2].

Despite that, there are still many unresolved issues which have not reached full agreement; however, public discussion can never solve all the problems and bring satisfaction to all. It should be directed toward the truth, even if it eventually reaches only compromise.

Today, there is a great tendency from upper level for another dialogue between science and religion, which existed since the very beginnings of our culture. Religion existed before science, but science is not an extension of religion. Each of them must keep their principles, their different interpretations, and their own conclusions. Although different, both are components of a common culture of humanity.

One of the most controversial topics in modern bioethics, science, and philosophy is the beginning of individual human life. In the seemingly endless debate, strongly stimulated by recent technologic advances in human reproduction, a synthesis between scientific data and hypothesis, philosophical thought, and issues of humanities has become necessary to deal with ethical, juridical, and social problems [3]. Furthermore, in this field, there is a temptation to ask science to choose between opinions and beliefs that tend to neutralize one another. Indeed, the question of when human life begins requires the essential aid of different forms of knowledge. Here we become involved in the juncture between science and religion, which needs to be carefully explored [4].

Obviously, the beginning of human life is seen differently by different individuals, groups, cultures, and religions. Fundamental to productive debate and reconciliation between minority and majority groups is an understanding of the ill-defined concept of "the beginning of human life" [5].

Entering this field, scientists have been remiss in failing to translate science into the terms that allow mankind to share their excitement of discovering life before birth. Regardless of the remarkable scientific development, curiosity, and speculations dating back to Hippocrates, life before birth still remains a big secret. Different kinds of intellectuals involved themselves in trying to contribute to the solution of the human life puzzle. They are led by the idea that each newborn child will only reach its full potential if its development in utero is free from any adverse influence, providing the best possible environment for the embryo/fetus. Considering the embryo/fetus, it should be always kept in mind the amazing aspect of these parts of human life in which the pregnant woman and the embryo/fetus, although locked in the most intimate of relationships, are at all times two separate individuals. Accepting the embryo/fetus as a person opens a new set of questions about its personality and human rights.

The Definition of Life

Proper answers to the question of how to define human life are complicated. Nowadays, dilemmas consider the respect of human life from the birth to death involving not just biology, but other sciences also. Philosophy, theology, psychology, sociology, law, and politics evaluate this topic from different point of views. Integration of all could result in a useful answer.

Some authors say that life as such does not exist—no one has ever seen it. Szent-Gyorgy says that the noun "life" has no significance because there is no such thing as "life." Le Dantez holds that the expression "to live" is too general and that it is better to say a dog "dogs" or a fish "fishes" than a dog or a fish lives [6].

When defining life, it should be considered not just as it is today, but as it might have been in its primordial form and as it will be in the future. All present forms of life appear as something completely new. Life, then, is transferred and not conceived in each new generation. Furthermore, the phenomenon of life has existed on Earth for approximately 3.5 billion years. Consequently, although the genome of a new embryo is unique, the make-up of an embryo is not new. If life is observed through the cell, then every life (and human also) is considered as a continuum. Human cells and mankind have existed on Earth continuously since the appearance of the first man. However, if the definition refers to a single human being or the present population, the statement that "human life is a continuum" is not acceptable [7, 8].

Life, in a true sense of the word, begins when the chemical matter gives rise, in a specific way, to an autonomous, self-regulating, and self-reproducing system. Life is connected with a living being, and it creates its own system as an indivisible whole—it forms its individuality. One of the most important characteristics of living beings is reproduction. Reproduction is a means of creating new life by transferring forms of an old one into newly formed human being. Therefore, variability, individual development, and harmony characterize human beings. Individuality is the most essential characteristic of human beings consisting of new life, but also all human life forms through evolution, characterized by phenotype, behavior, and the capability to recognize and adapt. Human embryo and fetus gradually develop into these characteristics.

"Human life" poses a semantic problem. The placenta is "human life," as is every individual cell or organ of the human body, but "human life" is clearly not equivalent to "human being." It is, therefore, mandatory to differentiate between organic or vegetative human life and "potential personal human life." The latter term allows various groups to identify a point of the continuum between abortion and birth to which they can ascribe appropriate values and rights [5].

Although we should not forget that in the same way today's research is tomorrow's benefit [6], concerning human life, conclusions should not be treated one-sidedly from one perspective. This reality should be regarded in all its richness: the embryo gives the biologist and geneticist substance for consideration, but talking about the beginning of a human life requires philosophical/anthropological consideration, as well as theological and social sciences. In its protection, we have to include ethics and law. This approach leads to the conclusion that it is necessary to reject reductionism as well as integrism and to find a "golden middle" between these two methodologies [3].

What Does Biology Say?

Biology characterizes human beings by the dynamics of the system and its self-control (homeostasis), excitability (response to stimuli of different nature and origins), self-reproducibility, the heredity of the characters, and the evolutionary trend [3]. For biologists, it is important to specify which form of life phenomena we are referring to: cell, organism population, or species. The basic level of organization and the simplest form of life is the cell. Biologically speaking, human cellular life never stops, or if it did, the extinction of the human species would result and is passed on from one generation to another. Human individual organismic life is defined within its life cycle, which is temporarily limited, i.e., it has a beginning and an end [9]. It is obvious that life is a highly dynamic phenomenon that could be described and explained through the careful study of life processes and interactions by interdisciplinary approach. In human spermatozoa and oocyte are two essential cells involved in creating human life. It is clear that biologists are most qualified to render judgment on the structure and function of cells. To quote Scarpelli [10], the very broad scope of biological science (from molecular to behavioral biology and from unicellular to multisystem forms) brings with it the justifiable understanding that the biological scientist knows and is able to define the state of being alive or "life." If not, the science fails.

The biological scientist, who may specialize within one or another domain of the broad scope, has particular and definitive knowledge and understanding of the living individual that is his specialty. If not, disorder will rise above failure.

Understanding of the beginning of human life and development of the embryo/fetus could provide definitive resolution. However, with the recent possibility of visualizing early human development virtually from conception, perinatologists should be those who by study, training, practice, and research are singularly qualified [11].

While science provides us data about physical development of the human being, it does not provide information about its personality and personhood. These are philosophical, rather than scientific, topics.

Human Embryogenesis

Only proper understanding of the process of human embryogenesis enables answering scientifically the question of when the life cycle of a human individual starts. Therefore, in the following text, the main steps of the human developmental process are going to be briefly described, primarily during the first 15 days following fertilization.

A human being originates from two living cells, the oocyte and the spermatozoon, transmitting the torch of life to the next generation. The oocyte is a cell approximately 120 μm in diameter with a thick membrane, known as the zona pellucida. The spermatozoon moves, using the flagellum or tail, and the total length of the spermatozoon including the tail is 60 μm [12].

After syngamy, the zygote undergoes mitotic cell division as it moves down the fallopian tube toward the uterus. A series of mitotic divisions then leads to the development of the preembryo. The newly divided cells are called blastomeres. From 1 to 3 days after syngamy, there is a division into two cells and then four cells. Blastomeres form cellular aggregates of distinct, totipotent, undifferentiated cells that, during several early cell divisions, retain the capacity to develop independently into normal preembryos. As the blastocyst is in the process of attaching to the uterine wall, the cells increase in number and organize into two layers of cells. Implantation progresses as the outer cell layer of the blastocyst, the trophectoderm, invades the uterine wall and erodes blood vessels and glands. Having begun 5 or more days after fertilization with the attachment of the blastocyst to the endometrial lining of the uterus, implantation is completed when the blastocyst is fully embedded in the endometrium several days later. Even during these 5–6 days, modern medicine introduces the possibility of making preimplantation genetic diagnosis.

However, at this time, these cells are not yet totally differentiated in terms of their determination to specific cells or organs of the embryo. The term preembryo, then, includes the developmental stages from the first cell division of the zygote through the morula and the blastocyst. By approximately the 14th day after the end of the process of fertilization, all cells, depending on their position, will have become parts of the placenta and membranes or the embryo. The embryo stage, therefore, begins approximately 16 days after the beginning of the fertilization process and continues until the end of 8 weeks after fertilization, when organogenesis is complete [13].

The preembryo is the structure that exists from the end of the process of fertilization until the appearance of a single primitive streak. Until the completion of implantation, the preembryo is capable of dividing into multiple entities, but does not contain enough genetic information to develop into an embryo; it lacks of genetic material from maternal mitochondria and of maternal and parental genetic messages in the form of messenger RNA or proteins. Therefore, during the preembryonic period, it has not yet been determined with certainty that a biological individual will result or would be one or more (identical twins forming), so that the assignment of the full rights of an individual human person is inconsistent with biological reality.

One conclusion from this is that the preembryo requires the establishment of special rules in the society: it cannot claim absolute protection based on claims of personhood; although meriting respect, it does not have the same moral value that a human person has. Today, one largely accepted opinion is that until the 14th day from fertilization or at least, until implantation—the human embryo may not be considered, from the ontological point of view, as an individual.

Genetic uniqueness and singleness coincide only after implantation and restriction have completed, which is about 3 weeks after fertilization. Until that period, the zygote and its sequelae are in a fluid process, are not physical individual, and, therefore, cannot be a person.

It is well-known that high percentages of oocytes which have been penetrated never proceed on to further development and that many oocytes which do are thwarted so early in their development that their presence is not even recognized. It is suggested that 30% of conceptions detected by positive reactions to human chorionic gonadotropin (HCG) tests abort spontaneously before these pregnancies are clinically verified.

The newly conceived preembryo presents itself as a biologically defined reality. However, the status of the preembryo as an individual remains a great mystery. In the present scientific scene especially with the progress of ultrasound technologies, prenatal psychology and therapeutics opened a window into prenatal life of embryo and fetus confirming the evidence that the embryo/fetus is a true subject itself [14, 15].

Personality

Defining personality is very complex. There is still no clear definition of personality. One dictionary offers "what constitutes an individual as distinct person," but does not define what the "what" is. Another dictionary asserts "the state of existing as a thinking intelligent being." This definition might lead to the inference that personality increases pro rata with intelligence or that some people may not have a personality at all if we followed Bertrand Russell's dictum that "most people would rather die than think and many, in fact, do!" Kenneth Stallworthy's *Manual of Psychiatry* is more help with the definition that "personality is the individual as a whole with everything about him which makes him different from other people," because we can certainly distinguish fetuses from each other and from other people. With the next sentence—"personality is determined by what is born in the individual in the first place and by everything which subsequently happens to him in the second"—we are really in the field [3, 5].

Viewpoints on the nature of "personhood" and what it means ethically and legally vary widely. In his proposed Life Protection Act, Sass acknowledges that a fetus with formed synapses is not a "person" in the usual sense of the word, connoting consciousness and self-consciousness [16]. Veatch sees the problem as defining the life that has full moral standing [17], while Knutson [18] has noted that "those who employ spiritual or religious definitions of when life begins tend to place the beginning of life earlier than those who employ psychological, sociological, or cultural definitions."

Led by the truism, "No insignificant person was ever born," human beings should be valued from birth to natural death. It is hard to establish proper values and exact definitions. This becomes especially problematic when prenatal life is considered. The above stated truism opens an important question: "Is the person-unborn a person in the first place and, if so, is the person-unborn a 'significant' person?" [3].

Let us evaluate further present controversies. There is no doubt that the embryo and fetus in utero are biologically human individuals prior to birth. The child who is born is the same developing human individual that was in the mother's womb. Birth alone cannot confer natural personhood or human individuality. This is confirmed by preterm deliveries of babies who are as truly human and almost as viable as those whose gestation goes to full term. All the known evidence supports the human fetus being a true ontological human individual and consequently a human person in fact, if not in law. A human person cannot begin before the appropriate brain structures are developed that are capable of sustaining awareness. The same applies to a grossly malformed fetus. It would still be a human individual even if its human nature was not perfect or its functions quite normal. Nobody questions the humanity of a Down's syndrome fetus or child. A fetus or child with severe open spina bifida is not less of a human being. The same should be said for the live anencephalic fetus or infant with only brainstem functions. It is a human individual even if it lacks a complete brain and usually survives birth by only a few hours or a day.

"Person" and "personhood" are the legally operational terms in the United States and many other countries. Alternatively, "person" and "personhood" are replaced by terms such as "viable outside the uterus," "a woman's right to privacy," and "a woman's right to choose." In each case, viable, privacy, and choice, the life-support provider may legally order transfer of the dependent individual into a morbid environment. For this group, dilemma (which includes the stem cell, abortion, and cloning debates) is abated, but not resolved [5].

Human society created several standards in defining "person" or "human being" based on what is familiar and easy recognizable [3]. For example, a human speaks, understands, and laughs. Absence of these characteristics (mutism, autism, and stoicism) does not disqualify. To the contrary, the conclusion is that the characteristics we have come to associate with being a person may not be applicable to each individual person. Therefore, it is necessary to establish criteria for a definition of "person" in society and in time. Some prominent Italian professors [14] committed themselves to caring for the embryo in such a way, giving the same dignity to every patient, and the human conditions to grow and develop, to educate others inside and outside the specialty, and to carry out research involving all the components of society.

Embryo as a Patient

Bioethical Aspects

The idea of the embryo/fetus as a miniaturized infant or adult is true to the extent that the embryonic/fetal physiologist must be able to apply knowledge of every system after birth, yet quite untrue in failing to recognize the many ways in which life before birth differs fundamentally from life after birth [6]. The newly conceived form presents itself as the biologically defined reality: it is an individual that is completely human in development that autonomously, moment by moment without any discontinuity, actualizes its proper form in order to realize through intrinsic activity, a design present in its own genome [14]. The embryo as a patient is best understood as the subset of the concept of the fetus as the patient. These two concepts opened a whole set of questions regarding ethical problems. The embryo as the patient is indivisible from its mother. However, balance is needed in protecting the interests of the embryo/fetus and the mother. One prominent approach to understanding the concept of the embryo/fetus as a patient has involved attempts to show whether or not the embryo/fetus has independent moral status or personhood [19–21]. Independent moral status for the fetus would mean that one or more of the characteristics possessed either in or of the embryo/fetus itself, and therefore independently of the pregnant woman or any other factor, generate and therefore ground obligations to the embryo/fetus on the part of the pregnant woman and her physician.

A wide range of intrinsic characteristics has been considered for this role, e.g., moment of conception, implantation, central nervous system development, quickening, and the moment of birth [22]. Given the variability of proposed characteristics, there are many views about when the embryo/fetus does or does not acquire independent moral status. Some take the view that the embryo/fetus possesses independent moral status from the moment of conception or implantation. Others believe that the embryo/fetus acquires independent moral status in degrees, thus resulting in "graded" moral status. Still others hold, at least implicitly, that the embryo/fetus never has independent moral status so long as it is in utero [21].

Being a patient does not require that one possesses independent moral status [23]. Being a patient means that one can benefit from the application of the clinical skills of the physician [24]. Put more precisely, a human being without independent moral status is properly regarded as a patient when the following conditions are met: that a human being is presented to the physician for the purpose of applying clinical interventions that are reliably expected to be efficacious, in that they are reliably expected to result in a greater balance of goods over harms in the future of the human being in question [22]. In other words, an individual is considered a patient when a physician has beneficence-based ethical obligations to that individual.

To clarify the concept of the embryo/fetus as the patient, beneficence-based obligation is necessary to be provided. Beneficence-based obligations to the fetus and embryo exist when the fetus can later achieve independent moral status [24]. This leads to the conclusion that ethical significance of the unborn child is in direct link with the child to be born—the child, it can become.

Legal Status of the Embryo

When discussing law, it should be always kept in mind that medicine is international, but law is not. Before the era of Aristotle, who taught that human life begins when the fetus is formed, human life was considered to begin at birth. Prior to birth, the fetus was not an independent human being but, like an organ, part of the mother [25]. Thus the birth of a full-term infant has been used in the laws of various countries to signify the beginning of the human life that is to be protected.

Indeed, the status of the human embryo is not juridically defined and relies on the political, social, and religious influences in each country. Interestingly, nearly all countries of the Western world use the 12th week of pregnancy as the limit for legal abortion. It is not the end of the first trimester, which is 13.3 weeks, and there is no other particular biological event to justify this limit.

It is hard to answer the question when human life should be legally protected. At the time of conception? At the time of implantation? At the time of birth? In all countries (except Ireland and Liechtenstein), juridical considerations are based on Roman law. Roman civil law says that the fetus has rights when it is born or if it is born-nasciturus.

Few countries agree with the definition of the beginning of human personality at the time of conception. The majority does not grant legal status to the human embryo in vitro (i.e., during the 14 days after fertilization). Thus, even in the absence of legal rights, there is no denying that the embryo constitutes the beginning of human life, a member of the human family. Therefore, whatever the attitude, every country has to examine which practices are compatible with the respect of that dignity and the security of human genetic material [26].

Arguments for Beginning of Human Life and Human Person at Fertilization

The fundamental approaches of biomedical and social (secular) practice must begin with the understanding that the subject before birth is a person and that "personhood" is conferred by successful fertilization of the egg. To hide from this in silence or ignorance should be unacceptable to all, as stressed by Scarpelli [11].

The view that human life begins when sperm and eggs fuse to give rise to a single cell human zygote, whose genetic individuality and uniqueness remain unchanged during normal development, is widely supported. Because the zygote has the capacity to become an adult human individual, it is thought it must be one already. The same zygote organizes itself into an embryo, a fetus, a child, and an adult. By this account, the zygote is an actual human individual and not simply a potential one, in much the same way as an infant is an actual human person with potential to develop to maturity and not just a potential person. As Scarpelli pointed out, outside the realm of religious dogma, there has been no one whose existence can be traced back to any entity other than the fertilized egg. The biological line of existence of each individual, without exception, begins precisely when fertilization of the egg is successful [11].

The process of fertilization actually begins with conditioning of the spermatozoon in the male and female reproductive tracts. Thereafter, fertilization involves not only the egg itself, but also the various investments which surround the egg at the time it is released from the ovary follicle. Fertilization, therefore, is not an event, but a complex biochemical process requiring a minimum of 24 h to complete syngamy, that is, the formation of a diploid set of chromosomes. During this process, there is no commingling of maternal and paternal chromosomes within a single nuclear membrane (prezygote); after this process, the parental chromosome material is commingled (zygote).

Among the many other activities of this new cell, most important is the recognition of the new genome, which represents the principal information center for the development of the new human being and for all its further activities. For the better understanding of the very nature of the zygote, two main features are to be at least mentioned here. The first feature is that the zygote exists and operates from syngamy on as a being, ontologically one, and with a precise identity. The second feature is that the zygote is intrinsically oriented and determined to a definite development. Both identity and orientation are due essentially to the genetic information with which it is endowed. That is why many do believe that this cell represents the exact point in time and space where a new human individual organism initiates its own life cycle [3].

Arguments Against the Beginning of Human Life at Fertilization

Today, one largely accepted opinion is that until the 14th day from fertilization or at least until implantation, the human embryo may not be considered, from the ontological point of view, as an individual. There are at least five main reasons in favor of this opinion:

1. Before the formation of the embryonic disc, the embryo is “a mass of cells, genetically human,” “a cluster of distinct individual cells,” which are each “distinct ontological entities in simple contact with the others” [27]. The genetically unique, newly developed DNA, a genome, is not established until 48 h after sperm penetration. The ovum and sperm lie side by side for more than 48 h before they finally merge. In biological terms, this renders conception as a process that occurs overtime and not a specific point in time [5].
2. Until approximately the 14th day after fertilization, all that happens is simply a preparation of the protective and nutritional systems required for the future needs of the embryo. Only when the entity called embryonic disc is formed can the embryo develop into a fetus [28].
3. The monozygotic twins phenomenon or chimeras can occur. In fact, this seems to be the strongest reason why the embryo is denied the quality of individuality and as a proof that the zygote cannot be an ontologically human being. In approximately one-third of cases, the embryo divides at about the two cells stage, and in the other two-thirds, the inner cell mass divides within the blastocyst from day 38. Occasionally, the division takes place from day 8–12, but usually it is not complete, thereby forming conjoined identical twins or two-headed individuals. The chimera, resulting from the recombination of two individual to become one individuum (and detectable through genetic testing), provides another argument against the equivalence of conception and the beginning of human life: no individuum has died, yet one has ceased to exist.
4. Co-existence of the embryo with its mother is a necessary condition for an embryo belonging to the human species, and this condition can be obtained only at implantation [21]. However, there is evidence that development of a human embryo in vitro can continue well beyond the stage of implantation and that mouse embryos implanted under the male renal capsule can reach the fetal stage. It is also argued, or at least implied, that so many human embryos die before or after implantation that it would be lacking in realism to accept that the human individual begins before implantation.

   It is well-known that high percentages of oocytes which have been penetrated never proceed on to further development and that many oocytes, which do, are thwarted so early in their development that their presence is not even recognized. Up to 50% of ovulated eggs and zygotes recovered after operations were found so grossly abnormal that it would be very unlikely that they would result in viable pregnancies. It is also suggested that 30% of conceptions detected by positive reactions to human chorionic gonadotropin (HCG) tests abort spontaneously before these pregnancies are clinically verified. The scientific literature is not unanimous on the incidence of natural wastage prior to, and during, implantation in humans, varying from 15% to as much as 50%. The vast majority of these losses are due to chromosomal defects caused during gametogenesis and fertilization [29].

   Genetic uniqueness and singleness coincide only after implantation and restriction have completed, which is about 3 weeks after fertilization. Until that period, the zygote and its sequelae are in a fluid process and are not a physical individual and therefore cannot be a person. Although in a set of twins, one individuum can disappear, genetic and individual identities are now more or less equivalent. Many eminent Catholic writers, among them the Australian priest Norman Ford, author of *When Did I Begin?*, consider implantation to mark the beginning of human life; they maintain that the preembryo has only intrinsic potential and must be protected only from the time of implantation [30].

5. The product of fertilization may be a tumor, a hydatidiform mole, or chorioepithelioma. Though the mole is alive and of human origin, it is definitely not a human individual or human being. It lacks a true human nature from the start and has no natural potential to begin human development.

   A teratoma is another clear instance of cells developing abnormally that results from the product of fertilization, but which could not be considered to be a true human individual with a human nature. It has no potential to develop into an entire fetus or infant. Clearly, the fetus with the teratoma would be a human individual, but not the attached teratoma itself. Obviously, not all the living cells that develop from the conceptus, the early embryo, or the fetus form an integral part of a developing human individual [3].

Different Religious Teachings and Historical Aspects

The Catholic Church's teachings are clearly described in the Introduction Donum Vitae: "A human creature is to be respected and treated as a person from conception and therefore from that same time his (her) rights as a person must be recognized, among which in the first place is the invaluable right to life of each innocent human creature."

In 1997, the Third Assembly of the Pontifical Academy for Life was held in Vatican City. It has been concluded that "at the fusion of two gametes, a new real human individual initiates its own existence, or life cycle, during which—given all the necessary and sufficient conditions—it will autonomously realize all the potentialities with which he is intrinsically endowed." The embryo, therefore, from the time the gametes fuse, is a real human individual, not a potential human individual. It was even added that recent findings of human biological science recognize that in zygotes resulting from fertilization, the biological identity of a new human individual is already constituted [31, 32].

In Western Europe and in North and South America, these opinions are mostly based on Judeo-Christian theology; in Arabian Countries, in Africa, and in Asia prevail the influences of the Islamic and Buddhist religions. Although their approach to the beginning of human life is impressively similar, each of these religions has different attitudes to the problem of embryo research, infertility, and its therapy. In a fact, while the Jewish attitude toward infertility is expressed in the Talmud sayings and in the Bible (synthesized in the first commandment of God to Adam "Be fruitful and multiply"), the Christian point of view establishes no absolute right to parenthood. According to the Islamic views, attempts to cure infertility are not only permissible but also a duty.

Islamic teaching is based on prophet Mohammed description: "The creation of each of you in his mother's abdomen assumes a 'nufta' (male and female semen drops) for 40 days, then becomes 'alaga' for the same (duration), then a 'mudgha' (like a chewed piece of meat) for the same, then God sends an angel to it with instructions. The angel is ordered to write the Sustenance, life span, deeds and whether eventually his lot is happiness or misery, then to blow the Spirit into him" (Human developments as described in Khur'an and Sunnah; Moore, et al. In: Some evidence for the truth of Islam, 1981). The summary of this poetic and sacred description is as follows: soul breathing "ensoulment" occurs at 120 days of gestation from conception.

To make this religious principle applicable to the practice, the Islamic Jurisprudence Council wrote a Fatwa in 1990 that said: "Abortion is allowed in the first 120 days of conception if it is proven beyond doubt that the fetus is affected with a severe malformation that is not amenable to therapy, and if his life, after being born, will be a means of misery to both him and his family, and his parents agree" so that there is no difficulty either for the prenatal diagnosis or for the possible termination of pregnancy within the exposed limits.

Buddhism has imposed strict ethics on priests, but it has relatively lenient attitudes toward lay people, so if medical treatment for infertility is available, people should make use of it.

For about 2000 years, the opinions of Aristotle, the great Greek philosopher and naturalist, on the beginning of the human being were commonly held. He argued that the male semen had a special power residing in it, pneuma, to transform the menstrual blood, first into a living being with a vegetative soul after 7 days and subsequently into one with a sensitive soul 40 days after contact with the male semen [33].

Aquinas adopted Aristotle's theory, but specified that rational ensoulment took place through the creative act of God to transform the living creature into a human being once it had acquired a sensitive soul. The first conception took place over 7 days, while the second conception, or complete formation of the living individual with a complete human nature, lasted 40 days [34].

Hippocrates believed that entrance of the soul into the male embryo occurred on the 30th day of intrauterine life. It entered into the female embryo on the 40th day. Actually, this idea was a considerable improvement on the scheme found in the Book of Leviticus, where it is suggested that the soul does not enter the female until 40 days after the conception [35].

In short, the rational soul enables the matter to become a human being, an animated body, an embodied soul, a human person.

Harvey's experiments with deer in 1633 proved Aristotle's theory of human reproduction wrong, without himself finding a satisfactory explanation of human conception. After modern scientists discovered the process of fertilization, most people took for granted that human beings, complete with a rational soul, began once fertilization had taken place.

It is clear that the answer to the question "When has the human being actually come to life?" could only be given by combining the cognition of different religions, philosophies, and various biological scientific disciplines. There is a very fine line between the competence of science and the one of metaphysics, and it greatly depends on the individual's philosophical principles. Those two, more or less autonomous intellectual disciplines, have very often tried dominating one another or ignoring each other. It is only recently that the majority of scientists and some theologians have come to realize that the separate meanings of scientific and religious "truths" complement each other, thus representing methodologically independent entities. Current science is not interested in what nature is, but in the facts that could be stated regarding it, thus trying to explain the term, rather than inventing it. The main difference between science and religion can be seen in the fact that scientific "truths," unlike religious postulates, can and must be experimentally verified, and the methods of scientific cognition can be easily explained and learnt. Whereas religion favors irrationality, science prefers an entirely rational approach to matters of importance. Intellectual cognition, when scientifically expressed, usually is in a form of mathematical formulas and presented quantitatively. Contrarily, religion tends to keep its truths in a form of metaphoric expressions, preferring qualitative. Today, there is a tendency, on a higher level, to reopen the dialogue between the science and religion, which was present at the very beginning of our culture. Religion had existed long before science came to life, but science is not to be thought of as a continuation of the religion. Each discipline should preserve its principles, its separate interpretations, and its own conclusions. In the end, both of them represent different components of the one and indivisible culture of mankind.

Clinical Controversies

There are some clinical controversies pertinent in any discussion of when life begins. Spermatozoa are living cells. They present evidence that they are living by their motility. They are equipped with an effective mechanism for movement in the form of a tail that beats under the control of the cytoplasmic droplets within the head. These living cells, which have been manufactured in the testes, are released into the environment provided by the male reproductive tract. They are not yet capable of fertilization. The spermatozoon must first come under the influence of the male reproductive tract, where it acquires the ability to function in fertilization. Even after ejaculation, it is capable of penetrating the egg, and it is modified further by exposure to the female reproductive tract, taking on the ability or capacity to fertilize. The decision must be made as to whether the spermatozoon is a being (i.e., living and human with the potential for continued life once fertilization has occurred); albeit in another form, it is entitled to the right of protection as a person. Those who deny right for life to the spermatozoon might argue that it is not a complete human cell chromosomally—it contains only the haploid number of chromosomes. Paradoxically, those who take that point of view would insist that an individual born with fewer or more chromosomes than normal is human and entitled to all the rights of "personhood." As Mastroianni stressed, the decision to base the definition of "human life" solely on the number of chromosomes in a given cell has far-reaching implications [36].

Furthermore, life has been defined as being terminated when brain activity ends. If we were to say that life begins when brain activity starts, we would be admitting that the definition of the beginning of life is dependent upon technology and not upon ethics or morality.

Some suggested that the beginning of human life requires the neural fusion of the periphery with the center, as well as sufficient development of the brain itself [37]. Brody formulated the so-called symmetry concept: if the death of a human being requires the death of the brain, the beginning of human life shall correspond with the beginning of the life of the brain, considered to be at day 32 pc [38]. However, Sass has correctly pointed out that fusion is not established anatomically without neurons which form synapses, which would be expected from embryological development at 70 days (8 weeks) pc [39].

In this light, let us take for example the accepted definition of birth, which some years ago was described as the complete expulsion of a fetus of 1000 g or 28 weeks of pregnancy. With advances in perinatal and neonatal intensive care, the line was drawn at 500 g, or approximately 22 weeks of gestation, some years later. This meant that a 20-week-old fetus was not born by definition, even if it was viable. This concept has changed. The same logic applies to a live fetus being accorded the term "life," if we use such definitions as the beginning of brain activity or ultrasonic proof of heartbeat and movement. The establishment of each of these parameters is shifted to an earlier stage year by year by improving technological refinements in electronic and ultrasonic equipment. This leads us to the conclusion that to follow this line of reasoning means to give life, birth, and viability definitions determined by technology. The more advanced the technology, the earlier life begins.

In any consideration of the beginning of human life, it helps to think about when life ends. Let us consider the following: a 2-week-old newborn is hospitalized with massive brain injury suffered in an automobile accident. Despite all measures, no electrical or other brain activity can be detected during the next 2 days, and the child is pronounced dead. Its body parts may survive after its death, as after the death of every person of whatever age. Hair and nails grow for days. Kidneys, heart, liver, and other organs may go on living for years if transplanted into another individual. Cells taken soon after death and cultured in a laboratory might live well beyond the 72 or more years this child might have lived, although the life of the infant has ended. The conclusion reached in this case that death of the brain means the end of life is generally accepted by physicians, courts, and the public [6].

Returning to the question of when life begins, it is true that the DNA of the fertilized egg has the information necessary to form an individual, but so does virtually every other cell in the body. Nobody would claim full rights for the living cells of the infant killed in the accident, although each has a complete library of DNA. Nor would they for thousands of living skin cells we lose every time we wash our hands and faces. Is there some stage in the development of the brain that is critical? Or is it the time at which the fetus can survive outside the womb, with or without the support of medical technology? Should we revert to a criterion used for many years, the time of quickening, when one can feel the fetus moving? These are questions still to be answered.

Visualization of Early Human Development

Significant advances have been made in recent years in visualizing and analyzing the earliest human development. Most of them have been done by introduction of three-dimensional static and color Doppler and 4D sonography. Many new parameters about early human development are now studied directly by new ultrasound techniques.

Considerable number of biochemical, morphological, and vascular changes occur within the follicle during the process of ovulation and luteinization, and most of them can be studied by transvaginal ultrasound with color Doppler and 3D facilities [40]. If the oocyte is fertilized, the embryo is transported into the uterus where under favorable hormonal and environmental conditions, it will implant and develop into a new and unique individual. The introduction of transvaginal color Doppler improved the recognition of blood vessels enabling detailed examination of small vessels such as arteries supplying preovulatory follicle, corpus luteum and endometrium [28].

Perifollicular vascularization can help in identification of follicles containing high-quality oocytes, with a high probability of recuperating, fertilizing, cleaving, and implanting, while 3D ultrasound enables accurate morphological inspection and detection of cumulus oophorus. Follicles without visualization of the cumulus by multiplanar imaging are not likely to contain fertilizable oocytes. This information is especially useful in patients undergoing ovulation induction.

Following ovulation, the corpus luteum is formed as the result of many structural, functional, and vascular changes in the former follicular wall. Color Doppler studies of the luteal blood flow velocities enable evaluation of the corpus luteum function in second phase of menstrual cycle and early pregnancy. When the placenta takes over the role of production of progesterone, the corpus luteum starts regressing.

After ovulation, there is a short period during which the endometrial receptivity is maximal. During these few days, a blastocyst can attach to the endometrium and provoke increased vascular permeability and vasodilatation at the implantation site. Trophoblast-produced proteolytic enzymes cause the penetration of the uterine mucosa and erode adjacent maternal capillaries. This results in formation of the intercommunicating lacunar network—the intervillous space of the placenta. A small intradecidual gestational sac can be visualized by transvaginal sonography between 32 and 34 days [41].

The secondary yolk sac is the earliest extraembryonic structure normally seen within the gestational sac in the beginning of the fifth gestational week. The yolk sac volume was found to increase from 5 to 10 weeks' gestation. When the yolk sac reaches its maximum volume at around 10 weeks, it has already started to degenerate, which can be indirectly proved by a significant reduction in visualization rates of the yolk sac vascularity [27]. Therefore, a combination of functional and volumetric studies by 3D power Doppler helps to identify some of the most important moments in early human development.

The embryonic heart begins beating on about day 22–23, accepting blood components from the yolk sac and pushing blood into the circulation. The embryonic blood begins circulating at the end of the 4th week of development.

The start of the embryo-chorionic circulation changes the source of nourishment to all intraembryonic tissues. The survival and further development of the embryo become dependent on the circulation of embryonic/fetal blood. If the embryo-chorionic circulation does not develop, or fails, the conceptus is aborted. The embryo cannot survive without the chorion (placenta), and the chorion will not survive without the embryo. Avascular degenerated chorionic villi constitute the hydatidiform mole.

Within the embryo, there are three distinct blood circulatory systems [12]

1. Vitelline circulation (from yolk sac to embryo).
2. Intraembryonic circulation.
3. Two umbilical arteries (from embryo to placenta-fetoplacental circulation).

It is possible to visualize and assess them virtually from conception [42–46].

At 5 weeks from the maternal side of placenta, it is possible to obtain simultaneous three-dimensional imaging of the developing intervillous circulation during the first trimester of pregnancy. Three-dimensional power Doppler reveals intensive vascular activity surrounding the chorionic shell starting from the first sonographic evidence of the developing pregnancy during the fifth week of gestation.

At 7 weeks, three-dimensional power Doppler images depict aortic and umbilical blood flow. Initial branches of umbilical vessels are visible at the placental umbilical insertion.

During the 8th–9th week, developing intestine is being herniated into the proximal umbilical cord.

At 9–10 weeks, herniation of the mid-gut is present. The arms with elbow and legs with knee are clearly visible, while feet can be seen approaching the midline.

At 11 weeks, three-dimensional power Doppler imaging allows visualization of the entire fetal and placental circulation.

During the 11th–12th week of pregnancy, development of the head and neck continues. Facial details such as nose, orbits, maxilla, and mandibles are often visible. Herniated mid-gut returns into the abdominal cavity.

New Possibilities for Studying Embryonic Movements and Behavior

The latest development of 3D and 4D sonography enables precise study of embryonic and fetal activity and behavior (Fig. 4.1) [47]. With four-dimensional ultrasound, movements of head, body, and all four limbs and extremities can be seen simultaneously in three dimensions [48]. Therefore, the earliest phases of the human

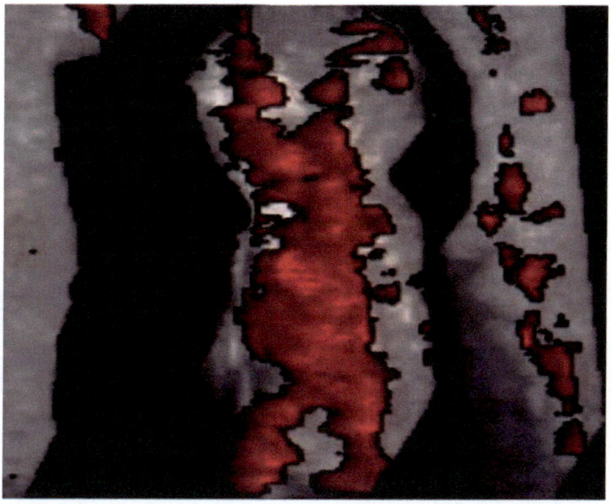

Fig. 4.1 Early triplets clearly visualized by three-dimensional sonography

anatomical and motor development can be visualized and studied simultaneously (Fig. 4.1). It is clear that neurologic development—early fetal motor activity and behavior—needs to be re-evaluated by this new technique [49–51]. Our group studied the development of the complexity of spontaneous embryonic and fetal movements [52]. With the advancing of the gestational age, the movements become more and more complex. The increase in the number of axodendritic and axosomatic synapses between 8 and 10, and again between 12 and 15 weeks [53], correlates with the periods of fetal movement differentiation and with the onset of general movements and complex activity patterns, such as swallowing, stretching, and yawning, seen easily by 4D technique. By 7–8 weeks of pregnancy, gross body movements appear. They consist of changing the position of the head toward the body. By 9–10 weeks of pregnancy, limb movements appear. They consist of changing the position of the extremities toward the body without the extension or flexion in the elbow and knee. At 10–12 weeks of pregnancy, complex limb movements appear. They consist of changes in the position of limb segments toward each other, such as extension and flexion in the elbow and knee.

Between 12 and 15 weeks of pregnancy, swallowing, stretching, and yawning activities appear. In addition to these activities, it is now feasible to study by 4D ultrasound a full range of facial expression including smiling, crying, and eyelid movement.

It is hoped that the new 4D technique will help us have a better understanding of both the somatic and motoric development of the early embryo. It will also enable the reliable study of fetal and even parental behavior [48].

There were recently a number of papers on new attractive techniques for visualization of early human development [54–66] (Figs. 4.2, 4.3 and 4.4).

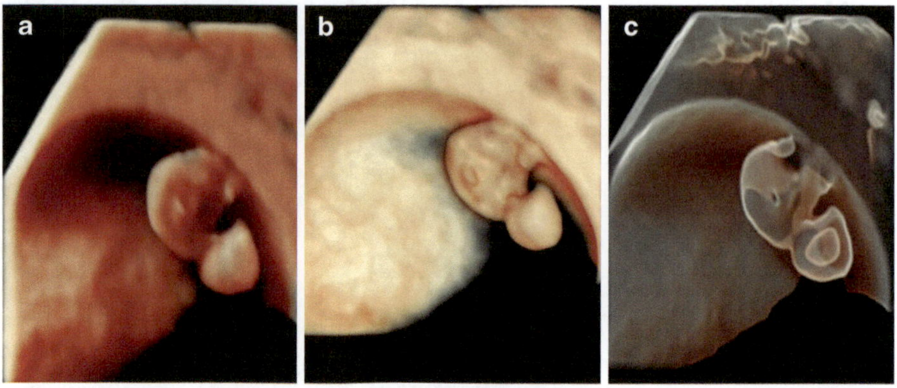

Fig. 4.2 (**a**–**c**) Six weeks HDlive silhouette images. (**a**) Conventional HDlive image of embryo and yolk sac. (**b**, **c**) With gradual increase of silhouette

Fig. 4.3 Six weeks HDlive flow image of maternal–embryonal circulation

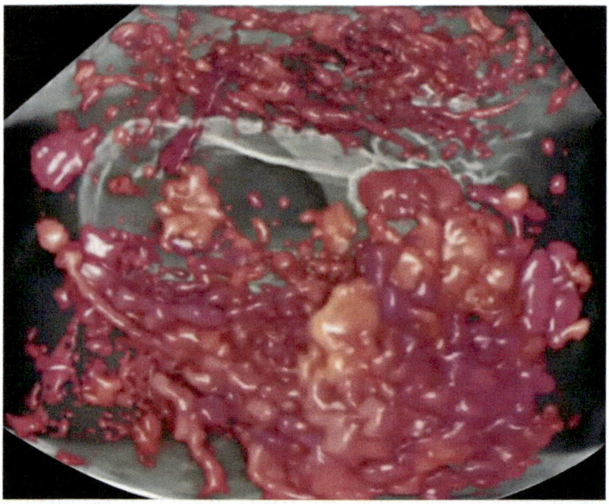

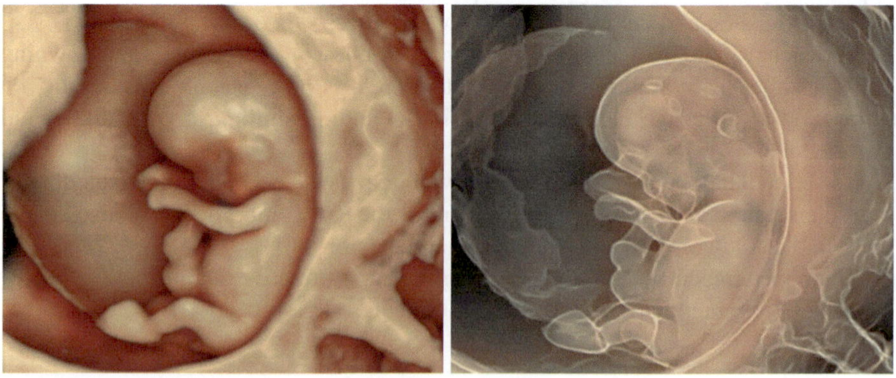

Fig. 4.4 Ten weeks HDlive silhouette image of embryo and amnion

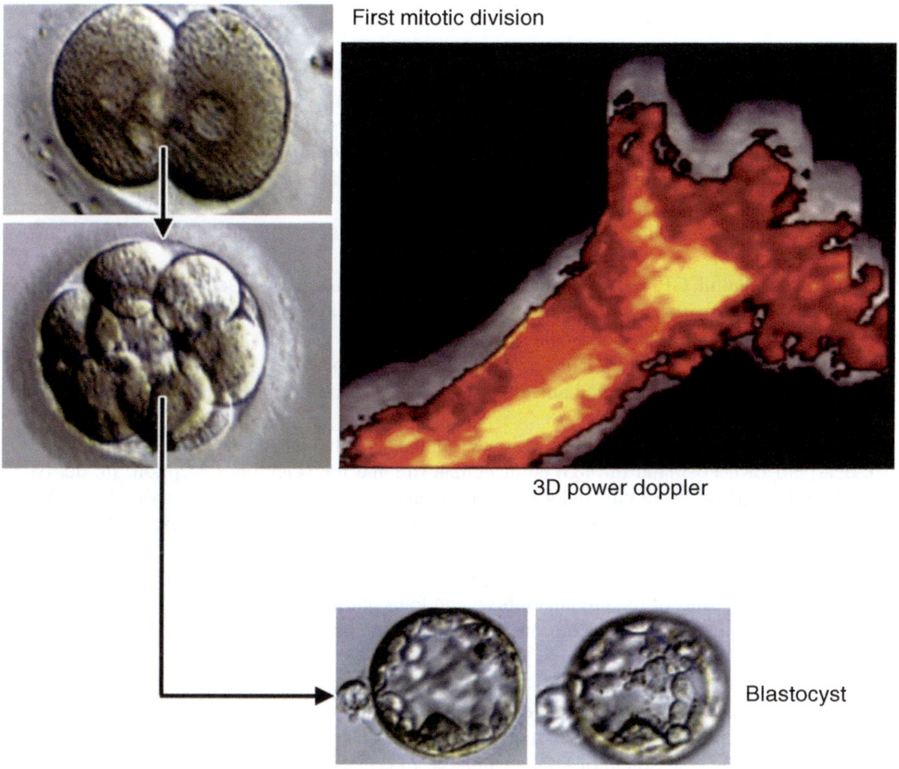

Fig. 4.5 Visualization of the patency of fallopian tube by three-dimensional power Doppler sonography, important for successful first mitotic division and transfer of early embryo to uterus

Conclusion

The question of when a human life begins and how to define it could be answered only through the interconnecting pathways of history, philosophy, medical science, and religion (Fig. 4.5). It has not been easy to determine where to draw the fine line between the competence of science and metaphysics in this delicate philosophical field. To a large extent, the drawing of this line depends on one's fundamental philosophical outlook. To quote Beller: "The point at which human life begins will always be seen differently by different individuals, groups, cultures, and religious faiths. In democracy, there are always at least two sides, and the center holds only when the majority realizes that without a minority, democracy itself is lost. The minority in turn must realize its best chance lies in persuasion by reason and thoughtfulness rather than fanaticism" [5].

References

1. Rolston H. Science, religion, and the future. In: Richardson WM, Wildman WJ, editors. Religion and science history, method, dialogue. New York: Routledge; 1996. p. 21–82.
2. Mayr E. The growth of biological thought. Harvard University Press: Cambridge; 1982. p. 81.
3. Serra A, Colombo R. Identity and status of the human embryo: the contribution of biology. In: de Dios Correa V, Sgreccia E, editors. Identity and statute of human embryo. Vatican City: Libreria Editrice Vaticana; 1998. p. 128.
4. Kurjak A. The beginning of human life and its modern scientific assessment. Clin Perinatol. 2003;30:27–44.
5. Beller FK, Zlatnik GP. The beginning of human life. J Assist Reprod Genet. 1995;12(8):477–83.
6. Kurjak A. When does human life begin? Encyclopedia Moderna. Cambridge: Cambridge University Press; 1992. p. 383–90.
7. Godfrey J. The pope and the ontogeny of persons (commentary). Nature. 1995;273:100.
8. Graham L. Foreword. In: Nathanielsz PW, editor. Life before birth and the time to be born. Ithaca: Promethean Press; 1992.
9. Gilbert SF. Developmental biology. Sunderland: Mass Sinauer Associates; 1991. p. 3.
10. Scarpelli EM. Postnatal through adult human life and the scientific deception. In: del Atti I, editor. Congresso Nazionale della Societa Italiana di Medicina Materno Fetale, Rome, March 17–21, 2003. Medimond (International Proceedings). Bologna: Monduzzi Editore; 2003. p. 29–36.
11. Scarpelli EM. Personhood: a biological phenomenon. J Perinat Med. 2001;29:417–26.
12. Jirasek JE. An atlas of the human embryo and fetus. New York: Parthenon Publishing; 2001.
13. ACOG Committee Opinion. Committee on ethics: preembryo research: history scientific background, and ethical considerations. Int J Gynecol Obstet. 1994;45:291–301.
14. Declaration of Professors from Five Faculties of Medicine and Surgery of the universities of Rome, organizers of the Conference: The Embryo as a Patient.
15. Kurjak A, Stanojevic M, Azumendi G, Carrera JM. The potential of four-dimensional ultrasonography in the assessment of fetal awareness. J Perinat Med. 2005;33:46–53.
16. Sass HM. Brain life and brain death: a proposal for normative agreement. J Med Philos. 1989;14:45–59.
17. Veatch RM. The beginning of full moral standing. In: Beller FK, Weir RF, editors. The beginning of human life. Dordrecht: Kluwer; 1994. p. 19.
18. Knutson AL. When does human life begin? Viewpoints of public health professionals. Am J Publ Health. 1967;57:2167.
19. Engelhardt HT Jr. The foundation of bioethics. New York: Oxford University Press; 1986.
20. Dunstan GR. The moral status of the human embryo. A tradition recalled. J Med Ethics. 1984;10:38–44.
21. Chervenak FA, McCullough LB, Kurjak A. Ethical implications of the embryo as a patient. In: Kurjak A, Chervenak FA, Carrera JM, editors. The embryo as a patient. New York: Parthenon Publishing Group; 2001. p. 226–30.
22. Curran CE. Abortion: contemporary debate in philosophical and religious ethics. In: Reich WT, editor. Encyclopedia of bioethics. New York: Macmillan; 1978. p. 17–26.
23. Ruddick W, Wilcox W. Operating on the fetus. Hastings Cen Rep. 1982;12:10–4.
24. McCullogh LB, Chervenak FA. Ethics in obstetrics and gynecology. New York: Oxford University Press; 1994.
25. Connery JR Jr. The ancients and medievals on abortion. In: Horan DJ, Grant ER, Cunningham PC, editors. Abortion and constitution. Washington, DC: Georgetown University Press; 1987. p. 124.
26. Pierre F, Soutoul JH. Medical and legal complications. J Gynecol Obstet Biol Reprod (Paris). 1994;23(5):516–9.
27. Ford NM. When did I begin? Conception of the human individual in history, philosophy and science. Cambridge: Cambridge University Press; 1991. p. 137–46.

28. McLaren A. Prelude to embryogenesis, in the Ciba Foundation, human embryo research, yes or no? vol. 12. London: Tavistock; 1986. p. 5–23.
29. Abel F. Nascita e morte dell'uomo: prospective della biologia e della medicina. In: Biolo S, editor. Nascita e morte dell'uomo. Problemi filosofici e scientifici della bioetica. Genoa: Marieti; 1993. p. 37–53.
30. McCormick KA. Who or what is the preembryo? Kennedy Instit Ethics J. 1991;1:24.
31. Mahoney SJ. Bioethics and belief. London: Sheed and Ward; 1984. p. 80.
32. Johnson M, Delayed hominization. Reflections on some recent Catholic claims for delayed hominization. Theol Stud. 1995;56:743–63.
33. Congregation for the Doctrine of the Faith. Instruction on respect for human life in its origin and on the dignity of procreation "Donum vitae" (February 12, 1987). Acta Apostolicae Sedis. 1988;80:70–102.
34. Ford NM. When did I begin? Cambridge: Cambridge University Press; 1991.
35. Beazley JM. Fetal assessment from conception to birth. In: Kurjak A, editor. Recent advances in ultrasound diagnosis. Amsterdam: Excerpta Medica; 1980. p. 128.
36. Kurjak A. Kada pocinje zivot. In: Kurjak A, editor. Ocekujuci novorodjence. Zagreb: Naprijed; 1987. p. 18–28.
37. Mastroianni L Jr. Ethical aspects of fetal therapy and experimentation. In: Schenker JG, Weinstein D, editors. The intrauterine life: management and therapy. Amsterdam: Excerpta Medica; 1986. p. 3–10.
38. Beller FK, Reeve J. Brain life and brain death: the anencephalic as an explanatory example. J Med Philos. 1989;14:5–20.
39. Brody B. Abortion and the sanctity of human life: a philosophical view. Cambridge: MIT Press; 1975. p. 109.
40. Sass HM. The moral significance of brain-life-criteria. In: Beller FK, Weir RF, editors. The beginning of human life. Dordrecht: Kluwer; 1994. p. 57–70.
41. Kupesic S. The first three weeks assessed by transvaginal color Doppler. J Perinat Med. 1996;24:310–7.
42. Kupesic S, Kurjak A, Ivancic-Kosuta M. Volume and vascularity of the yolk sac. J Perinat Med. 1999;27:91–6.
43. Kurjak A, Predanic M, Kupesic S. Transvaginal color Doppler study of middle cerebral artery blood flow in early normal and abnormal pregnancy. Ultrasound Obstet Gynecol. 1992;2:424–8.
44. Kurjak A, Kupesic S. Doppler assessment of the intervillous blood flow in normal and abnormal early pregnancy. Obstet Gynecol. 1997;89:252–6.
45. Kurjak A, Kupesic S, Hafner T. Intervillous blood flow in normal and abnormal early pregnancy. Croatian Med J. 1998;39(1):10.
46. Kurjak A, Kupesic S. Three-dimensional transvaginal ultrasound improves measurement of nuchal translucency. J Perinat Med. 1999;27:97–102.
47. Kurjak A, Kupesic S, Banovic I, Hafner T, Kos M. The study of morphology and circulation of early embryo by three-dimensional ultrasound and power Doppler. J Perinat Med. 1999;27:145–57.
48. Lee A. Four-dimensional ultrasound in prenatal diagnosis: leading edge in imaging technology. Ultrasound Rev Obstet Gynecol. 2001;1:144–8.
49. Campbell S. 4D, or not 4D: that is the question. Ultrasound Obstet Gynecol. 2002;19:1–4.
50. de Vries JI, Visser GH, Prechtl HF. The emergence of fetal behaviour. I. Qualitative aspects. Early Hum Dev. 1982;7:301–22.
51. de Vries JI, Visser GH, Prechtl HF. The emergence of fetal behaviour. I. Qualitative aspects. Early Hum Dev. 1985;12:99–120.
52. Kurjak A, Vecek N, Hafner T, Bozek T, Funduk-Kurjak B, Ujevic B. Prenatal diagnosis: what does four-dimensional ultrasound add? J Perinat Med. 2002;30:5762.
53. Kurjak A, Azumendi G, Vecek N, Kupesic S, Solak M, Varga D, et al. Fetal hand movements and facial expression in normal pregnancy studied by four-dimensional sonography. J Perinat Med. 2003;31:496–508.

54. Kurjak A, Azumendi G, Andonotopo W, Salihagic-Kadic A. Three- and four-dimensional ultrasonography for the structural and functional evaluation of the fetal face. Am J Obstet Gynecol. 2007;196:16–28.
55. Kurjak A, Miskovic B, Andonotopo W, Stanojevic M, Azumendi G, Vrcic H. How useful is 3D and 4D ultrasound in perinatal medicine? J Perinat Med. 2007;35:10–27.
56. Kurjak A, Miskovic B, Stanojevic M, Amiel Tison C, Ahmed B, Azumendi G, Vasilj O, Andonotopo W, Turudic T, Salihagic-Kadic A. New scoring system for fetal neurobehavior assessed by three- and four-dimensional sonography. J Perinat Med. 2008;36(1):73–81.
57. Pooh RK, Kurjak A. Recent advances in 3D assessment of various fetal anomalies. Donald Sch J Ultrasound Obstetr Gynecol. 2009;3(3):1–25.
58. Pooh RK, Shiota K, Kurjak A. Imaging of the human embryo with magnetic resonance imaging microscopy and high-resolution transvaginal 3-dimensional sonography: human embryology in the 21st century. Am J Obstet Gynecol. 2011;204(77):e1–16.
59. Pooh RK, Kurjak A. Three-dimensional/four-dimensional sonography moved prenatal diagnosis of fetal anomalies from the second to the first trimester of pregnancy. Donald Sch J Ultrasound Obstetr Gynecol. 2012;6(4):376–90.
60. Pooh RK, Kurjak A. Novel application of three-dimensional HDlive imaging in prenatal diagnosis from the first trimester. J Perinat Med. 2014;43(2):147–58.
61. Kurjak A, Spalldi Barisic L, Stanojevic M, Salihagic Kadic A, Porovic S. Are we ready to investigate cognitive function of fetal brain? The role of advanced four-dimensional sonography. Donald Sch J Ultrasound Obstetr Gynecol. 2016;10(2):116–24.
62. Kurjak A, Spalldi BL, Delic T, Porovic S, Stanojevic M. Facts and doubts about the beginning of human life and personality. Donald Sch J Ultrasound Obstetr Gynecol. 2016;10(3):205–13.
63. Kurjak A. Editorial 3D/4D sonography. J Perinat Med. 2017;45(6):639–41.
64. Pooh RK, Kurjak A. Donald School atlas of advanced ultrasound in obstetrics and gynecology. New Delhi: Jaypee Brothers; 2015.
65. Kurjak A, Spalldi BL, Ahmed B, Porović S, Hasanović A. Beginning of human life: science and religion closer and closer. In: Kurjak A, Chervenak FA, McCullough LB, editors. Hasanović a: science and religion—synergy not Skepticism. Jaypee Brothers: New Delhi; 2018. p. 21–52.
66. Kurjak A, Spalldi BL. Controversies on the beginning of human life—science and religion closer and closer. Psychiatr Danubina. 2021;33(3):S257–79.

5 Philosophical Considerations About the Beginning of Humana Life and Their Clinical Implications

Frank A. Chervenak and Amos Grunebaum

Introduction

Philosophical considerations about the beginnings of human life have important clinical implications for obstetricians and gynecologists not only for daily clinical applications but also for basic science and clinical research in reproductive medicine.

Obstetrician and gynecologists already are familiar with ethical considerations, especially about the moral status of the embryo and the fetus but are less familiar with philosophical considerations that underlie these familiar ethical considerations.

Metaphysics is the branch of philosophy that studies the fundamental nature of reality, the first principles of being, identity and change, space and time, causality, necessity, and possibility.

Understanding philosophical and metaphysical considerations and their connections to the ethical considerations is essential for obstetricians and gynecologists but also other specialties concerning professionally responsible clinical practice and research. Here, we provide a concise and accessible introduction to the metaphysics of human reproduction, its philosophical reasoning, and metaphysics and its implications for ethical reasoning. We also will discuss the connections to the ethics of human reproduction and then to the implications of these philosophical considerations for the professional ethics of research and clinical practice in obstetrics and gynecology.

F. A. Chervenak (✉) · A. Grunebaum
Department of Obstetrics and Gynecology, Zucker School of Medicine and Lenox Hill Hospital, New York, NY, USA
e-mail: fchervenak@northwell.edu

J. G. Schenker et al. (eds.), *Hot Topics in Human Reproduction*, Reproductive Medicine for Clinicians 3, https://doi.org/10.1007/978-3-031-24903-7_5

Philosophical Reasoning

Any philosophical reasoning comprises of at least two steps [1]:

1. The first step comprises the effort to become as clear as possible about concepts pertinent to the topic at hand.
2. The second step is to identify the implications of these concepts for the topic at hand.

Essential Concept

The first type of concept is known as an *essential concept. An essential concept* can be stated as a set of criteria for the invocation of the concept, more precisely, the individually necessary and jointly sufficient conditions for the invocation of a concept. The individually necessary conditions are:

(a) A justified claim, a claim for which reasons can be given that others are intellectually obligated to accept.
(b) Treatment by others, what they should or should not do to the rights-bearer.
(c) The specific behavior that counts as acceptable treatment by others.

That each of these three is an individually necessary condition means that, if any one of them is not satisfied, the concept of a right cannot justifiably be invoked. That these three are jointly sufficient conditions means that, if all three are satisfied, then the concept of a right can justifiably be invoked.

Cluster Concept

The second type of concept is more complex and is known as a *cluster concept.* A cluster concept differs from an essential concept in that not all of the sufficient conditions have to apply in every case; some are sufficient to invoke the concept. For a cluster concept, different groupings of criteria can serve as sufficient conditions. When they are satisfied, the cluster concept can be invoked. These groupings can differ in different contexts, which puts cluster concepts at risk of becoming unclear. The antidote to this risk is the clear statement of the pertinent sufficient conditions and why they are satisfied.

Obstetrician–gynecologists, other physicians, and biologists are already well acquainted with a cluster concept, which is the focus of this article: life. Britannica (better known as the Encyclopedia Britannica) defines life: "Life, living matter and, as such, matter that shows certain attributes that include responsiveness, growth, metabolism, energy transformation, and reproduction" [2]. The Oxford English Dictionary provides the following definition of life: "The condition that distinguishes animals and plants from inorganic matter, including the capacity for growth, reproduction, functional activity, and continual change preceding death" [3]. Other sources provide similar definitions. These definitions and others demonstrate that the concept of life is a cluster concept.

Inorganic matter is not considered "life" as it meets none of the criteria in any definition of life: the criteria specified in a definition of life are jointly necessary conditions for the invocation of the concept. Inorganic matter is lifeless, although its antecedents may have been alive, as in the case of fossils.

Life can exist by satisfying some criteria but not necessarily all. For example, life can exist without reproduction. When a post-menopausal patient is not capable of reproduction, she can still be very much alive. The patient retains the capacities for functional activity, energy transformation, growth, and continued change. But when none of the necessary conditions for being alive are satisfied, then there is no life but absence of life or death.

Life can display variation in complexity, from single-cell organisms to bodily organs, to the body as a whole, and in large eco-systems. There are also borderline, challenging cases, including viruses and prions.

This brief review of the cluster concept of life underscores a point we made above: when a cluster concept is deployed, the criteria deployed to invoke the concept must be clearly stated and justified as pertinent and biologically grounded. Failure to do so will always result in a paralyzing lack of clarity, in which circumstance philosophical reasoning is impossible.

Particulars and Individuals

Metaphysics can be a dense subject to study, sometimes impenetrably dense, which is not acceptable in philosophy and therefore not acceptable in professional ethics in obstetrics and gynecology. The word "metaphysics" originated in the need for a title to an early edition of the works of Aristotle for a text that had no title but appeared after his text, "Physics" [4]. The untitled text became known as "Metaphysics" or, from the Ancient Greek, "After the Physics." In the history of Western philosophy, metaphysics has become to be understood as the study of the most fundamental aspects of being or reality.

An important subset of metaphysics is known as ontology or the classification of entities or a typology of entities. Ontology is a branch of metaphysics concerned with the nature and relations of being. Aristotle's text, *Categories*, for example, is one of the earliest works in ontology in the history of Western philosophy. For Aristotle, there are two basic categories: substances or individuals of various natural kinds and their properties, which Aristotle called "accidents" to indicate that, while the nature of a substance does not change, its properties do [5]. Like Aristotle, for reproductive medicine and research, we need an account of two types of entities, particulars and individuals, which have precise meanings in ontology.

Particulars

The concept of a particular is an essential concept and contains a single criterion that functions therefore as both the necessary and the sufficient condition for invoking the concept: the entity in question can be distinguished from other entities. For example, a cell in one petri dish in a laboratory can be distinguished from a cell in another petri dish. This is known as spatial difference. One cell can divide into two, with the latter coming after the former in time. This is known as temporal difference. Spatial and temporal differences are the most common form of differences invoked in metaphysics to distinguish particulars. The criterion of distinguishability is known in the history of metaphysics as the criterion of distinction [6–9].

Individuals

The concept of an individual is an essential concept that comprises two criteria: distinction and indivisibility into two entities of the same kind [6–9]. A patient in an obstetrician–gynecologist's clinic is an individual. She can be distinguished spatially from the patients in the other examination rooms. She can be distinguished temporally from patients who preceded and who will follow her in the examination room. She is also indivisible: she is not capable of dividing into two human beings. If it becomes clinically justified to amputate one of her lower extremities to surgically manage gangrene, the result is not two human beings but one human being now without a portion of one of her lower extremities and the severed extremity. The severed extremity can be divided, so it is a particular not an individual.

In the history of Western metaphysics in the Aristotelian tradition, the source of both distinction (spatially or temporally distinguished) and indivisibility is known as the principle of individuation. This is a constitutive component of individuals. There are differing accounts of the principle of individuation, but this need not concern us for present purposes. The key point is that all accounts agree that the principle of individuation generates both distinction and indivisibility [6–9].

Ethical Reasoning About Particulars and Individuals

Ethical reasoning is defined as a form of reasoning whose behavior is right and good. In the history of moral philosophy, many differing ethical theories exist about what should count as right and good. There is agreement, however, that when we have an obligation to protect and promote the interests of an entity, then that entity has what is called "moral status."

Moral Status

Moral status means that there are good reasons, a justification, for such obligations which are called ethical obligations or sometimes also moral obligations [1]. There are two kinds of moral status [1]: the "dependent moral status" and the "independent moral status."

Dependent Moral Status

The first moral status, the dependent moral status, is a moral status that we attribute to an entity because we have an interest in it or a stake in its present and future existence. For example, a couple hoping to initiate a pregnancy using embryos produced by in vitro fertilization have a stake in the present and future existence of the in vitro embryos, especially those that have been evaluated to be good candidates for transfer. Dependent moral status is given by others to an entity and does not originate in some aspect of that entity.

Independent Moral Status

The second kind of moral status is called independent moral status. This means that we have obligations to an entity because it has the capacity to generate its own moral status independent of the interests of others. In other words, independent moral status is an essential concept in which the capacity to generate moral status functions as the necessary and sufficient condition for such generation. An entity has independent moral status if (the sufficient condition) and only if (the necessary condition) it has the capacity.

There are competing accounts in the history of Western philosophy and also in other global philosophical traditions about what capacity is required as the criterion for self-generated moral status. Some accounts emphasize that the independent moral status of a human being is a function of having a central nervous system that supports consciousness that includes both sensory awareness and self-awareness. Other accounts emphasize that sensory awareness and consciousness sufficient to experience pain (a report in the central nervous system of tissue damage or threat of tissue damage accompanied by awareness) generates independent moral status. In the more than 2500-year history of Western philosophy, there is no agreement in the global history of philosophy about which account of independent moral status must be accepted by all. Any claim, therefore, to have established *the authoritative* account of independent moral status lacks philosophical validity [1].

Metaphysical reasoning and ethical reasoning about particular and individuals and their moral status have two major implications for the ethics of reproductive medicine and research:

1. *Only biological individuals have the capacity that generates independent moral status, whatever that capacity may be according to a particular ethical theory.*
2. *Biological particulars are not capable of having the capacity to generate independent moral status.*

Biologic individuals with independent moral status have this status independently of the interests of others. This has the important ethical implication that everyone must acknowledge independent moral status. Biologic particulars do not have this capacity because they can divide or twin. Particulars, therefore, can only have dependent moral status. Biologic particulars with dependent moral status have this status solely as a function of the interests of others. Being a particular organism *and* being the object of the interests of others are the individually necessary and jointly sufficient conditions for having dependent moral status.

The Metaphysics of Human Reproduction

Gametes or sex cells are an organism's reproductive cells. There are usually female and male gametes. Ova or egg cells are female gametes, and sperms or spermatozoa are male gametes. Gametes are haploid cells, which means that each cell or gamete carries only one copy of each chromosome.

Fertilization begins when an egg or ovum joins with one or more sperm, leading to cell division and eventually creating an embryo. Gametes thus have the capacity to fuse to become an embryo. Because individual cells can divide, the constituent cells of an embryo are particulars. These blastomeres form a coherent group of cells known as a morula. The morula has the capacity to twin, so it is also a particular. Even as the morula becomes progressively complex in its organization, it remains a particular. The embryo that retains the potential to divide into twins does not satisfy the criterion of indivisibility but only the criterion of distinction. Such embryos are not individuals but particulars. It follows that the in vitro embryo is a particular not an individual. This is also the case for the in vivo embryo before it implants in the uterine wall and no longer has the capacity for twinning. When this capacity is lost, the implanted embryo and the fetus that it becomes are now individuals because they satisfy the two criteria of distinction and indivisibility.

Hurlbut has taken the view that embryos are indeed individuals in virtue of "an unbroken continuity in the differentiation and organization of the emerging individual life," and he is almost alone in taking on the metaphysical challenges of early human life forms [10, 11]. The problem for Hurlbut's assumption is that the most that "unbroken continuity" of a coherent set of dividing cells can establish is distinction. Hurlbut is aware of the challenge of twining just described but attempts to sidestep this challenge by claiming that twinning is the result of "a disruption of normal development by a mechanical or biochemical disturbance of fragile cell relationships" [10]. Such a disruption would not be possible for an organism that had achieved indivisibility, a metaphysical constraint on scientific explanation that Hurlbut does not acknowledge. He then claims that, before twinning, there exists a "crucial relational dynamics of position and intercellular communication are already at work establishing the unified pattern of the emerging individual" [10]. The concept of an emerging individual is left unexplained, which is not consistent with the requirement of clarity in philosophical reasoning. The claim that there is somehow an emerging individual in the coherent collection of cells does not defeat the metaphysical analysis that, before twinning becomes impossible, this collection is a particular and not an individual.

As the Human Embryo Research Panel of the US National Institutes of Health put it, "developmental individuation" is only achieved after twining becomes impossible [12]. Before this occurs, embryological development is not, as Hurlbut would have it, "unified" but, instead, highly coherent. The highly coherent nature of the pre-implantation embryo is a source of its distinction from other embryos, e.g., in a petri dish in a reproductive embryology laboratory. In addition, on Hurlbut's account of the embryo as a unified organism, obtaining a single cell for pre-implantation genetic/genomic diagnosis would result routinely in destruction of the embryo, which turns out not to be the case. This scientific reality confirms that the pre-implantation embryo is a particular, not an individual.

Implications of the Metaphysics of Human Reproductions for the Ethics of Human Reproduction

Because they are particulars and not individuals, gametes, in vitro embryos, and in vivo embryos before implantation do not have independent moral status, but they have only dependent moral status. Any claim that organisms that are particulars, but not individuals, possess independent moral status is philosophically invalid.

In vivo embryos that have implanted in the uterine wall and are growing as well as fetuses are individuals. However, being an individual is not enough to establish that they have independent moral status. In other words, being an individual is a necessary condition for having independent moral status: being an individual in and by itself is not a sufficient condition for having independent moral status. To have independent moral status, biologic individuals, organisms that are distinct and indivisible, must also satisfy an additional necessary and sufficient condition: the capacity to generate moral status independently of the interests of others, as explained above. In other words, when an individual organism has the capacity to generate its own moral status, it fulfills the sufficient condition for having independent moral status. Being an individual organism and having the capacity to generate moral status are the individually necessary and jointly sufficient conditions for having independent moral status.

The fetus does not have independent moral status because that would require a central nervous system that supports consciousness that includes both sensory awareness and self-awareness, which a fetus doesn't have [13]. The achievement of independent moral status comes only after birth, and there is disagreement in ethical theories about when after birth the central nervous system capacity to generate moral status exists.

Implications for Professionally Responsible Research and Practice in Reproductive Medicine

Gametes, embryos, and fetuses are continuums of life forms. They satisfy some of the sufficient conditions in the cluster concept of life, including responsiveness, growth, metabolism, energy transformation, and reproduction (in the form of cellular replication). The destruction of a living gamete, embryo, or fetus introduces a life-taking pathology not previously present and not interfering with that pathology as it runs its course to death.

That a biological entity was previously alive does not establish, by itself, whether its termination is permissible (shown in ethical reasoning to be acceptable) or impermissible (shown in ethical reasoning to be unacceptable). This ethical judgment can be made only on the basis of the moral status of a biologic organism. The current criteria for the cluster concept of life do not include the capacity to generate moral status (however it is understood in competing ethical theories). This is because the cluster concept of life in the science of evolutionary biology needs to be comprehensive, to include the full range of organisms, from the single-celled, particular organism to the complex, multi-celled, individual organism. That a biologic organism is alive cannot therefore *by itself* establish that that organism has either dependent or independent moral status.

It is well accepted globally that basic science and clinical research are required for the improvement of the safety and efficacy of patient care in obstetrics and gynecology and all other specialties [14]. Research into the beginnings of human life requires the use of gametes and pre-implantation embryos. In order for public policy (and consequently legislation) about the beginnings of life to have intellectual and moral authority in modern, pluralistic societies, public policy should be consistent with the requirements of philosophical reasoning about the beginnings of life as set out above. Public policy and legislation, specifically legislation that does not satisfy this requirement, will justifiably be considered arbitrary because it lacks intellectual and moral authority. Such public policy should not command respect but be expected to encourage cynicism and non-cooperation. These outcomes are perilous for the professional integrity of research on the beginnings of human life.

The requirement that public policy be consistent with philosophical reasoning has an important implication. The objection that such research is ethically impermissible on the grounds that gametes and pre-implantation embryos have independent moral status is philosophically invalid. Public policy about the regulation of such research should therefore not be based on any claim that gametes or embryos possess independent moral status. In particular, the claim that a pre-implantation embryo has independent moral status and therefore a right to life is philosophically invalid because the pre-implantation embryo does not satisfy the necessary condition of being a biological individual.

Many individuals and communities, especially faith communities, have an interest in the human organisms that constitute the beginnings of human life. This means that gametes and, especially, pre-implantation embryos are candidates for having dependent moral status because they are particulars not individuals. Policy makers immediately confront a problem: faith communities do not have the same interests in gametes and pre-implantation embryos. Some faith communities will find research on gametes and embryos objectionable or even impermissible, while other faith communities will find such research not only permissible but also obligatory.

The first step is to recognize that individuals and faith communities who object to this research experience the real moral burden of allowing such research to occur and to use public funds to pay for it. The incidence of infertility is not based on a patient's beliefs about the moral status of gametes and pre-implantation of embryos. This clinical reality means that those who object to basic science and clinical research to mitigate infertility might become candidates for clinical interventions based on such research. This means that individuals or groups in a pluralistic society must acknowledge that the moral burden they experience can be offset by the benefits of mitigations of infertility that might result from basic science and clinical research [15].

It follows that ethically justified public policy is to permit basic science and clinical research on gametes and pre-implantation embryos. Priority should be given to investigation into the mechanisms of infertility and how these might be safely altered to mitigate infertility.

Implications for the Professional Ethics of Clinical Practice of Reproductive Medicine

In the professional ethics of obstetrics and gynecology, practicing obstetrician–gynecologists have the professional responsibility to improve the safety and efficacy of clinical practice, including reproductive medicine. They can readily meet this ethical requirement by referring eligible patients to clinical research that has undergone prospective review and is approved by the legally designated entity. This is known in most institutions as an Institutional Review Board in the United States or Research Ethics Committee in other countries [14]. Such approved research will be normally based on sound science and ethical justification and will include an ethically appropriate informed consent process that respects science and the autonomy of their patients.

Some patients may decline the referral to this research and explain that they do so on moral grounds, especially moral grounds related to the moral status of gametes and pre-implantation embryos. To respect the autonomy of these patients and to show respect for them as persons, the obstetrician–gynecologist should share some thoughts that the patient may find worth considering before refusing the referral altogether. For patients who accept this offer, the obstetrician–gynecologist should set out the reasoning just above and its key point: the potential clinical benefits of mitigating infertility may offset the moral burden of research using gametes or pre-implantation embryos. The goal should be to fulfill the basic requirement of the ethical principle of respect for autonomy, empowering the patient to make an informed decision about whether to accept referral to a clinical trial [1].

Conclusion

Metaphysics may at first appear far removed from research and practice in reproductive medicine. In this article, we have shown that the latter and the former are intimately connected. A basic knowledge of the metaphysics of particulars and individuals helps to elucidate the concept of dependent and independent moral status and therefore the professional responsibilities of obstetrician–gynecologists regarding research and practice in reproductive medicine.

Funding Supported by Departmental funds.

Conflict of Interest The authors have no conflicts of interest to report.

References

1. McCullough LB, Coverdale JH, Chervenak FA. Professional ethics in obstetrics and gynecology. Cambridge: Cambridge University Press; 2020.
2. Life | Definition, Origin, Evolution, Diversity, and Facts | Britannica. https://www.britannica.com/science/life. Accessed 17 Mar 2022.
3. LIFE | Definition of LIFE by Oxford Dictionary on Lexico.com also meaning of LIFE. https://www.lexico.com/en/definition/life. Accessed 17 Mar 2022.
4. Smith JA, Ross WD. Aristotle, works of trans. into English under editorship of JA Smith and WD Ross. In: Barnes JA, editor. The complete works of Aristotle: the revised Oxford translation, vol. 1. Princeton: Princeton University Press; 1984. p. 3–24.
5. Ackrill J. Aristotle's categories. In: Barnes J, editor. The complete works of Aristotle: the revised Oxford translation, vol. 2. Princeton: Princeton University Press; 1984. p. 1552–728.
6. Gracia JE. Introduction to the problem of individuation in the early middle ages. Hanover: Philosophia Verlag; 1988.
7. Gracia JE. Individuality: an essay on the foundations of metaphysics. Albany: State University of New York Press; 1988.
8. Gracia JE. Individuation in scholasticism: the later middle ages and the counter-reformation 1150–1560. Albany: State University of New York Press; 1994.
9. McCullough LB. Leibniz on individuals and individuation: the persistence of premodern ideas in modern philosophy. Dordrecht: Kluwer Academic Publishers (now Springer); 1996.
10. Hurlbut WB. Framing the future: embryonic stem cells, ethics and the emerging era of developmental biology. Pediatr Res. 2006;59(4 Pt 2):4R–12R.
11. Khushf G. Owning up to our agendas: on the role and limits of science in debates about embryos and brain death. J Law Med Ethics. 2006;34(1):58–76.
12. Human Embryo Research Panel. National Institutes of Health, report of the human embryo research panel, vol. 1. Bethesda: National Institutes of Health; 1994.
13. Engelhardt HT Jr. The foundations of bioethics. 2nd ed. New York: Oxford University Press; 1996.
14. Brody BA. The ethics of biomedical research: an international perspective. New York: Oxford University Press; 1988.
15. Chervenak FA, McCullough LB. How physicians and scientists can respond responsibly and effectively to religiously based opposition to human embryonic stem cell research. Fertil Steril. 2008;90:2056–9.

Part III

Rights to Reproduce

ART: Right to Reproduce

6

R. A. Frankel, J. Merkison, and A. H. Decherney

ART: The Past and the Future

With the birth of Louise Brown on July 25, 1978—the first child successfully born as a result of in vitro fertilization (IVF)—came a rapid, international revolution in assisted reproductive technology (ART) unmatched by many other medical fields. Within 2 years of this milestone in the United Kingdom, the first US IVF clinic was opened in Virginia. Concurrently, the standardization of semen analyses and the early stages of preimplantation genetic diagnosis (PGD) were introduced, setting the stage for improved diagnostic techniques. Two years later, in 1982, the first babies born from frozen embryo transfer were delivered in Australia, as was the first delivery after a successful intrauterine insemination (IUI). Soon after, a woman with surgically absent ovaries underwent successful IVF using a donor oocyte, expanding the use of ART to women who would have otherwise been excluded. In 1984, the first baby conceived using a surrogate was born in California. In 1992, the process of intracytoplasmic sperm injection (ICSI) was successfully employed in Belgium and resulted in a viable pregnancy. In 2002, the first successful pregnancy after blastocyst biopsy using PGD was reported. In 2006, Louise Brown herself gave birth to a naturally conceived child, bringing full circle a whirlwind of technological advances in assisted reproduction [1].

R. A. Frankel
Department of Obstetrics and Gynecology, Donald and Barbara Zucker School of Medicine at Hofstra/Northwell, New York, USA
e-mail: rfrankel1@northwell.edu

J. Merkison · A. H. Decherney (✉)
Department of Reproductive Endocrinology and Infertility, Eunice Kennedy Shriver National Institute of Child Health and Human Development, National Institutes of Health, Bethesda, MD, USA
e-mail: decherna@mail.nih.gov

J. G. Schenker et al. (eds.), *Hot Topics in Human Reproduction*, Reproductive Medicine for Clinicians 3, https://doi.org/10.1007/978-3-031-24903-7_6

Advancements within each ART discipline allowed for improved treatment outcomes over time. For example, conversion in oocyte retrieval from a laparoscopic surgery to an ultrasound-guided procedure allowed this to become a primarily outpatient, same-day procedure [2]. In terms of diagnosis, PGD has undergone considerable transformation such that fewer cells can now be tested to predict hundreds of genetic disorders, improving overall diagnostic success [3]. Finally, techniques such as cryopreservation of both gametes and ovarian tissue have become pillars in fertility preservation, decreasing the pressure to transfer multiple embryos per cycle and, ultimately, the associated maternal and neonatal morbidity [4, 5].

In parallel with the fast-paced evolution of assisted reproduction came a proportional rise in public concern. The initial publication of IVF research in animal studies in the 1930s sparked ethical debate regarding the implication of reproductive freedom in humans [6]. Great concern from the government and public regarding the possibility of a "designer conception" or an "ideal race" tainted the innovative successes [7, 8]. Initially, falsified published reports claiming the use of ART for cloning caused public upheaval regarding the true intentions of ART [8]. More recently, controversial issues such as the impregnation of older women [9], the ownership over gametes or embryos [7], or sex selection via PGT [10] continue to spark debate. Even 44 years in these ethical and legal challenges add complexity to the scientific achievement.

Since 1978, more than five million births worldwide have resulted from ART [1, 2]. Studies project such exponential increase in the use of ART that, by the year 2100, nearly 167 million people will be borne by way of these interventions [11]. As a result, greater concern for societal order and ethical integrity will add important layers to the discussion and employment of assisted reproduction. The purpose of this chapter is to explore the pillars of present-day ART and the legal and ethical considerations they pose.

Pillars of Assisted Reproduction

Innovation within the field of assisted reproduction has provided previously infertile individuals and couples the opportunity to conceive in ways never before imagined [2, 12]. Before ART, diagnoses such as tubal or male factor infertility were considered insurmountable blockades. Early attempted reproductive procedures such as tubal reconstructive surgery, though promising, were largely unsuccessful and came with associated risk [12]. However, the basic science and clinical advancements in disciplines such as IVF, embryo transfer, ICSI, gamete and ovarian tissue cryopreservation, and preimplantation for genetic testing (PGT) have opened the possibility of parenthood for a wider pool of hopeful couples.

Conventional IVF involves the retrieval of oocytes from the ovaries with subsequent fertilization in a laboratory; the developing embryo is then transferred into the uterus at a later time for implantation. Even before the successful retrieval and transfer of the embryo that led to the birth of Louise Brown, the process of IVF has undergone many modifications. In its early stages, the oocyte retrieval process required laparoscopic surgery; while minimally invasive, this technique was still quite experimental, and unstimulated cycles often yielded few mature oocytes. As a result, the discovery of medications such as human menopausal gonadotropin (hMG) and gonadotropins to help control the menstrual cycle improved oocyte yield and overall success rates [12]. In 1983, the first successful birth occurred after the use of a donor oocyte with IVF, expanding the applicability of this technology to those with premature ovarian failure or advanced maternal age [12].

The development of gamete and embryo cryopreservation was the next step in allowing for fertility preservation for medical, personal, or research-oriented reasons. Specifically for patients undergoing gonadotoxic treatments, these advancements allowed for the conservation of future fertility that was previously unimaginable. Additionally, for those individuals or couples interested in delayed childbearing, cryopreservation superseded the natural obstacle of declining female fertility. The cryopreservation of semen was the first to be actualized due to the ease of obtaining semen samples as compared with the limited quantity of excess oocytes. In fact, oocyte cryopreservation was only perfected 30 years after spermatozoan cryopreservation was developed. Most importantly, studies have shown that the freeze-thaw process required for cryopreservation does not appear to impact the embryo viability or neonatal outcomes, making this a safe and important facet of ART [5].

Cryopreservation has provided women, in particular, the independence and autonomy to pursue reproduction at a personally satisfactory time, rather than that dictated by nature. It can often be safer for the patient, allowing for the preservation of embryos from a single stimulation cycle to be used in future cycles, thereby decreasing the simultaneous transfer of multiple embryos and the number of ovarian stimulation cycles a patient may need. Additional benefits include the ability to use PGT or delay embryo transfer for other medical reasons such as ovarian hyperstimulation syndrome (OHSS) or cancer diagnoses [5].

For prepubertal patients or those with hormone-sensitive cancers who cannot undergo ovarian stimulation for gamete cryopreservation, ovarian tissue cryopreservation provides an alternative route [13]. This process involves the biopsy of ovarian cortical tissue and primordial follicles or whole ovary excision followed by autologous reimplantation after completion of treatment [12, 14]. While this option confers hope for many people, ASRM cautions against its use for benign conditions or those desiring delayed childbearing [13].

Male factor infertility such as azoospermia or poor spermatozoa motility is one of the major causes of infertility in couples. The refinement of ICSI provided a bypass mechanism for fertilization that was not previously feasible. ICSI involves isolating a single sperm and directly injecting it into an oocyte in the laboratory setting. Over the following few days, fertilization occurs and an embryo develops. In contrast with conventional IVF, ICSI more reliably guarantees insemination. It also has proven benefit in cases of transmissible viral diseases, such as HIV, as it allows for removal of seminal fluid and the selection of a single spermatozoan to reduce the risk of transmission [15].

One discipline which has undergone tremendous advancement is that of PGD. PGD involves the aspiration of embryonic cells, which are analyzed for the presence of genetic disease [3]. Eventually this led to the development of preimplantation genetic testing for aneuploidies (PGT-A), which has the ability to differentiate genetically normal and varying degrees of abnormal embryos in an effort to prioritize transfer of embryos with the greatest pregnancy potential [3]. Given that chromosomal abnormalities increase proportionally with age, this technology has been most useful in cases of advanced maternal age, as well as recurrent implantation failure, recurrent pregnancy loss, and severe male factor infertility [3].

Legal Components of ART

With the rapid advances in reproductive technologies came a need for legal regulation that could quiesce growing public skepticism [8]. Both the maternal and neonatal morbidities and mortalities associated with ART, specifically those related to the risk of multiple gestations, gave rise for concern regarding the implications of unregulated use. As a result, a combination of federal, state, and professional organizations were developed to assist in reporting and maintaining standards within the practice [16].

In 1992, Congress passed the Fertility Clinic Success Rate and Certification Act (also known as the Wyden Law), which required structured reporting of clinic and cycle data to the Centers for Disease Control and Prevention (CDC) to made public. It allowed for more accessible information to both clinics and lay people regarding the prevalence of infertility diagnoses, procedures, and clinical pregnancy and live birth rates [17]. It also prompted the development of accrediting organizations which would regulate embryology laboratories [16, 18]. Additionally, the federal government mandated other agencies, such as the Food and Drug Administration (FDA) and the Centers for Medicare and Medicaid Services (CMS), to oversee various aspects of ART regulation [17].

The primary role of the FDA has been to oversee the development and use of infertility medications and devices. All new protocols or drug regimens are submitted to the FDA for approval prior to implementation [18]. The FDA also oversees the screening and storage of donor sperm or oocytes, to monitor the spread of infectious diseases and other necessary testing [17, 18].

Under the Clinical Laboratory Improvement Amendments (CLIA) of 1988, the primary role of the CMS became to ensure standardization and regulation of all laboratories which conduct testing on human specimens [17]. Responsibilities include coding and reimbursement, collection of fees, and clinic inspections [18]. This established quality assurance despite frequent use and testing of human gametes, embryos, and other tissues [17].

In the United States, individual states oversee their own clinics and physicians. Medical licenses are granted only to those who have met benchmarks and are expected to perform at a high medical and professional level. In cases of misconduct, the states are expected to legislate [17]. Insurance coverage policies are also mandated by the states, with some providing full coverage and others offering only partial or no coverage [19]. While this allows patients in full-coverage states the opportunity to obtain services needed, it also creates a notable access-to-care disparity for patients with less financial assistance [6, 11].

The development of professional organizations also assisted in the regulation and standardized reporting of ART clinics across the country. Two of the most well-known professional societies include the American Society for Reproductive Medicine (ASRM), established in 1944, and its affiliate the Society for Assisted Reproductive Technology (SART), founded in 1987 [18]. ASRM was developed to act both as a guidepost for clinics and physicians, as well as an advocate for patients through high-quality research shared with both the science and lay communities [20]. SART was developed specifically to collect data on clinics and publish public reports through the CDC [16]. They performed audits and site visits and included site information as well as cycle characteristics [8]. Participation was and continues to be voluntary; therefore, while this data is essential, it is not always comprehensive [8].

In addition to reporting, ASRM and SART also continue to publish guidelines and committee opinions on some of the legal and ethical issues associated with assisted reproduction [17, 18]. They set minimum standards for who should be running clinics and laboratories and how ART should be most safely performed [19]. For example, in 1999, ASRM published its first set of recommendations on the minimum number of embryos to be transferred in a single cycle, taking into account a woman's diagnosis and prognosis as well as the risks of multiple gestation; while this document has been edited over the years to reflect changes in technology and practice, it serves to standardize patient safety [16, 19, 21]. However, despite these comprehensive guidelines, ultimate decisions on how to practice are left up to the clinics and physicians [21].

The lack of federal and interstate consensus on ART regulation gives rise to considerable legal debate on how certain cases are to be handled. For example, the ownership and disposition of cryopreserved embryos is an important consideration in cases of excess embryos or inability of one partner to contribute to joint decision-making [22]. Another complex issue involves the concept of legal parenthood, particularly in cases of surrogacy in which the relationship between the surrogate and the intended parents or between the intended parents themselves has dissolved. While many clinics have their own policies and contracts regarding these potential situations, multiple court cases have arisen as a result of a lack of state standards [22]. The ability for ART to provide children to same-sex couples has also led to substantial debate, including the concept of parentage, insurance coverage, and discrimination of services by individual providers [22].

Ethical Components of ART

Alongside these ongoing legal conversations arose parallel ones regarding ethical considerations. New types of parenthood for single people, same-sex couples, or older women gave rise to a multitude of concerns. Some worried how children raised by older parents or those that don't fit heteronormative roles might fare [7, 9]. Arguments have been made that younger infertile woman should be prioritized over older women and that the worthiest donor oocyte recipients should be those with the most promising outcomes [9]. However, literature exploring these topics in depth have proven no difference in medical outcomes or psychological adjustment for children born into these atypical scenarios [7, 9].

The development of PGT has allowed for more successful ART outcomes such as decreased time to pregnancy and improved pregnancy per cycle rates [23]; it has also sparked great concern regarding the implications of genetic manipulation or sex selection [10, 23]. Some question the degree to which PGT should be used to exclude genetically unfavorable embryos and which diseases should ethically be considered "unfavorable" [7]. Additionally, the possibility of sex selection by way of PGT, especially in countries where one sex may be favored over another, has been highly debated and discouraged [7, 18].

Oocyte donation has also given rise to ethical concern. Firstly, women undergoing the process of oocyte donation must understand and be extensively counseled on the risks associated with this process, including OHSS and surgical complications [10, 23]. Some argue that the routine purchase of genetic material is an unethical practice and may lead to coercion or the possibility of discounted services in exchange for donated oocytes [7, 10, 23]. The incentivization of women to donate genetic material for anything other than altruistic motive is of great concern. Conversely, the question of how to handle excess embryos—be it indefinite storage, discarding, or donation to research or to other couples—has been the source of debate across various countries, states, religions, and communities [10].

Gamete cryopreservation has massively impacted family planning, particularly for women undergoing gonadotoxic therapy. While the practice provides security, arguments arise over ownership and disposition of gametes in cases of the patient's disability or death [23]. As with all aspects of ART, cryopreservation therefore requires thorough pretreatment counseling.

Access to care continues to be a major blockade for those pursuing ART due to the high cost of care [11]. While many states offer some form of insurance coverage, the out-of-pocket cost for most patients can be burdensome. As a result, patients and physicians may be inclined to make less favorable medical decisions (such as the transfer of multiple embryos) in the hopes of optimizing the chance of conception per cycle [6]. Alternatively, if associated cost poses too great a burden, it may cause personal debt or a premature cessation of treatment [6]. Therefore, the economic barriers to infertility treatment often worsen an already difficult situation for many patients and families.

Conclusion

The revolutionary era of reproductive medicine is nowhere near completed. With every great advance made in the field over the last 44 years came a litany of social and legal debate that challenged new technology and practice. As the field inevitably continues to progress, federal and state legislatures will undoubtedly continue to confront these issues with regulation, and society will continue to contend with the legal and ethical complexities that accompany it. As a result, it will be the physicians and professional societies who will have to continue advocating for patients, improving access to care, and ensuring the safe application of practice. In time, perhaps assisted reproduction will be seen as a necessary treatment for a common disease rather than a controversial or unethical aberrance to nature.

References

1. Kamel RM. Assisted reproductive technology after the birth of Louise Brown. J Reprod Infertil. 2013;14(3):96–109.
2. De Geyter C. Assisted reproductive technology: impact on society and need for surveillance. Best Pract Res Clin Endocrinol Metab. 2019;33(1):3–8.
3. Sciorio R, Tramontano L, Catt J. Preimplantation genetic diagnosis (PGD) and genetic testing for aneuploidy (PGT-A): status and future challenges. Gynecol Endocrinol. 2020;36(1):6–11.
4. Lawrenz B, Coughlan C, Melado L, Fatemi HM. The ART of frozen embryo transfer: back to nature! Gynecol Endocrinol. 2020;36(6):479–83.
5. Rodriguez-Wallberg KA, Waterstone M, Anastacio A. Ice age: cryopreservation in assisted reproduction—an update. Reprod Biol. 2019;19(2):119–26.
6. Ethics Committee of the American Society for Reproductive Medicine. Disparities in access to effective treatment for infertility in the United States: an Ethics Committee opinion. Fertil Steril. 2021;116(1):54–63.
7. Asplund K. Use of in vitro fertilization—ethical issues. Ups J Med Sci. 2020;125(2):192–9.
8. Chang WY, DeCherney AH. History of regulation of assisted reproductive technology (ART) in the USA: a work in progress. Hum Fertil (Camb). 2003;6(2):64–70.
9. Harrison BJ, Hilton TN, Riviere RN, Ferraro ZM, Deonandan R, Walker MC. Advanced maternal age: ethical and medical considerations for assisted reproductive technology. Int J Womens Health. 2017;9:561–70.
10. Brezina PR, Zhao Y. The ethical, legal, and social issues impacted by modern assisted reproductive technologies. Obstet Gynecol Int. 2012;2012:686253.
11. Faddy MJ, Gosden MD, Gosden RG. A demographic projection of the contribution of assisted reproductive technologies to world population growth. Reprod Biomed Online. 2018;36(4):455–8.
12. Wang J, Sauer MV. In vitro fertilization (IVF): a review of 3 decades of clinical innovation and technological advancement. Ther Clin Risk Manag. 2006;2(4):355–64.
13. Practice Committee of American Society for Reproductive Medicine. Ovarian tissue cryopreservation: a committee opinion. Fertil Steril. 2014;101(5):1237–43.
14. Rivas Leonel EC, Lucci CM, Amorim CA. Cryopreservation of human ovarian tissue: a review. Transfus Med Hemother. 2019;46(3):173–81.
15. O'Neill CL, Chow S, Rosenwaks Z, Palermo GD. Development of ICSI. Reproduction. 2018;156(1):F51–F8.

16. Rebar RW, DeCherney AH. Assisted reproductive technology in the United States. N Engl J Med. 2004;350(16):1603–4.
17. Medicine ASFR. Oversight of assisted reproductive technology. Washington, DC: ASRM; 2021.
18. Adamson D. Regulation of assisted reproductive technologies in the United States. Fertil Steril. 2002;78(5):932–42.
19. Hornstein MD. State of the ART: assisted reproductive technologies in the United States. Reprod Sci. 2016;23(12):1630–3.
20. Medicine ASFR. History of ASRM. Washington, DC: ASRM; 2020.
21. Frith L, Blyth E. Assisted reproductive technology in the USA: is more regulation needed? Reprod Biomed Online. 2014;29(4):516–23.
22. Casolo J, Curry-Ledbetter C, Edmonds M, Field G, O'Neill K, Poncia M. Assisted reproductive technologies. Georget J Gender Law. 2019;20(313):313–55.
23. Londra L, Wallach E, Zhao Y. Assisted reproduction: ethical and legal issues. Semin Fetal Neonatal Med. 2014;19(5):264–71.

Surrogate Pregnancies: Medical, Ethical, Legal, and Religious Aspects

7

Einat Gutman-Ido and Joseph G. Schenker

Types of Surrogacies

The surrogate mother is defined by the Council of the British Medical Association as "a woman who carries a fetus and bears a child on behalf of another person or persons, having agreed to surrender that child to this or these persons at birth or shortly thereafter" [1].

There are three forms of surrogacy, as follows [2, 3]

1. Partial natural surrogacy: This form has been known for thousands of years. The husband of the infertile woman has intercourse with another woman (the surrogate mother), who donates her genetic material and the use of her womb. The child is then given to the man who donated the sperm and to his legal wife, without adoption procedures.

   The first known surrogate mother was Hagar: "Now Sarai, Abram's wife bore him no children. And Sarai said unto Abram go unto my maid: it may be that I may obtain children by her. And Abram harkened to the voice of Sarai. And he went in unto Hagar and she conceived" [4, 5].

   Surrogacy of this form has been practiced by several tribes and cultures for thousands of years.
2. Partial surrogacy: The surrogate mother undergoes artificial insemination by sperm from the husband of the infertile woman. The oocyte and the womb are supplied by the surrogate, and the child born is then given to the father (donor of the sperm) and to his legal wife after legal adoption procedures. This method is practiced in several countries [6, 7] where surrogacy contracts are mediated by commercial agencies. In these cases, the surrogate mother is the genetic mother

E. Gutman-Ido (✉) · J. G. Schenker
Department of Obstetrics and Gynecology, Hebrew University-Hadassah Medical Center, Jerusalem, Israel

J. G. Schenker et al. (eds.), *Hot Topics in Human Reproduction*, Reproductive Medicine for Clinicians 3, https://doi.org/10.1007/978-3-031-24903-7_7

and the legal mother. The husband of the infertile woman, the sperm donor, is the genetic father, but may become the legal father only after adoption procedures. His wife is a social mother and becomes the legal mother after adoption.

3. Complete surrogacy [8–10]: This form is made possible only by the use of modern assisted reproductive techniques, such as IVF. The infertile woman donates the oocyte, and her husband donates the sperm. The pre-embryo, formed by in vitro fertilization, is placed in the womb of another woman, the surrogate mother.

   In practice, the relationship between the child and parents is established by the act of birth. The woman who carries the child is considered to be the legal mother, and the commissioning couple are not the legal parents. As a result of this, the genetic parents have to adopt the child after birth. The child is then given to the infertile couple after legal adoption, by which the genetic parents become the legal parents.

   In this type of surrogacy, the surrogate is a gestational mother only, providing the womb and not the genetic material.

   In 1985, the Mt. Sinai Clinic in Cleveland announced the first reported pregnancy of a surrogate gestational mother. A successful birth ensued in April 1986 [8].

Indications

There are several situations which may indicate the use of surrogacy

1. Uterine factors: congenital absence of the uterus or severe malformations of the Mullerian duct; severe Asherman syndrome; women after hysterectomy in the reproductive age; uterine leiomyomas; women who have been exposed to diethylstilbestrol.
2. Medical conditions which make pregnancy and delivery dangerous: severe cardiovascular disease; renal failure; severe hypertension; advanced collagen disease; severe diabetes.
3. Embryopathy uterine environment: maternal antibodies; maternal medications; maternal blood-borne diseases; inadequate nutrition.
4. Non-medical situations such as women who are not interested in pregnancy and delivery for reasons of social, professional, or mental psychological inconvenience.

Surrogacy: A Controversial Issue

Surrogacy is a source of great controversy in society, the medical profession, and the public. It is the basic right of every adult to marry and to procreate. On the other hand, it is not a state's obligation to provide new fertility techniques in order to enable procreation.

The World Health Organization (WHO) has stated that infertility is a state of disease. Nevertheless, the practice of surrogacy is not considered to be a medical treatment.

Arguments Against Surrogacy

The objections to surrogacy are based on the following

1. Surrogacy takes advantage of the gestational carrier who may be a woman of low socioeconomic class, from developing countries, or a family member under stress.
2. Surrogacy absolves the gestational carrier of responsibility.
3. Surrogacy can offend the gestational carrier's honor.
4. Surrogacy has commercial aspects.
5. There are medical, physical, and mental risks to the surrogate mother.
6. Surrogacy has been likened to prostitution in that it is an offence to a woman's dignity and tantamount to hiring out her female attributes.
7. Many argue that surrogate arrangements depersonalize reproduction and permit the separation between genetic, gestational, and social parenthood.
8. Professionals who attempt to serve both the couple and the surrogate, or who receive finder's fees for arranging surrogate relationships, may have a conflict of interests or exploit the parties.
9. Furthermore, the child is considered a product, and his best interests may be ignored.
10. Additionally, the ability to have one's genetic child without carrying the pregnancy raises the ethical issue of whether it is proper for a woman to contract with a surrogate gestational mother for social, rather than medical, reasons.

A woman who is physically capable of maintaining a pregnancy may wish to use a surrogate because she fears pregnancy or for reasons of convenience or vanity. In such cases, there is speculation that a woman's refusal to carry the pregnancy calls into question her ability to care for the child after its birth. It also raises the issue of whether it is ethical to ask a surrogate to undergo the risks of pregnancy when the genetic mother had no medical reasons to use a surrogate.

For the above reasons, some authorities in the field of IVF and embryo transfer (ET) feel that surrogacy should not be encouraged.

Arguments Supporting Surrogacy

The advocates of surrogacy base their case on the following statements [9]

1. It is the right of human beings to do whatever they please, as long as they do not harm another human beings.
2. Bearing children is the right of every person in society.
3. Surrogacy may enhance personal happiness.
4. Since surrogacy exists and will continue to do so despite legal limitations, it should not be prevented.
5. Studies have shown that most cases of surrogacy arrangements in Israel are carried out without conflict.

6. For some couples, the use of a surrogate gestational mother may be the only way of having their genetic child.
7. For the child, the use of a surrogate gestational mother gives him, or her, the opportunity that would otherwise not be available—the opportunity to exist.
8. The child, because it is so very much desired, is more likely to be cherished and brought up in a stable, loving environment, and this is to be encouraged.
9. The risk of exploitation of the surrogate is difficult to separate from the concept of her being "a laborer worthy of hire," which is performing gestational work in the reproductive process for payment of a fee. It is obvious that she carries both significant physical burden and risks. It is unreasonable to expect that any worker should undertake such risks without appropriate payment. This raises one of the controversial issues within surrogacy: should the surrogate be paid or be functioning out of purely altruistic reasons? Payment may appear to be a way of persuading women to choose to become surrogate mothers as a career option, something most authorities feel should not be encouraged.
10. The accepted fee in surrogate arrangements usually only covers expenses and cannot be regarded as a salary. This may serve to solve the moral issues involved in hiring without payment.

Statements by International Ethical Committees

Ethical committees in several countries have dealt with the issue of surrogacy.

Strong opposition to the surrogate program was expressed by the Warnock Committee in England:

"Legislation should be introduced... to render criminal the actions of professionals and others who knowingly assist in the establishment of a surrogate pregnancy... all surrogacy agreements are illegal contracts and therefore unenforceable in courts."

The Warnock Committee [10] recommended that commercial agencies should be outlawed. However, there is no prohibition of non-commercial agencies or individuals who might arrange a surrogacy, and there is no ban on payments being made to the surrogate mother to cover personal and medical expenses.

The FIGO Committee for the Study of Ethical Aspects of Human Reproduction has stated [11]: "The committee has strong reservations about surrogate practice, since it undermines the value of the family unit." Others claim that full surrogacy should be allowed under state regulations similar to those governing adoption, in order to find a solution for those infertility problems which can only be solved by surrogacy.

In the United States, the practice of surrogacy is not covered by federal law. States differ in their attitude to surrogacy. Some prohibit surrogacy contracts or make them void and unenforceable, while others permit such agreements [12].

There are a number of expert groups including the European Society of Human Reproduction and Embryology (ESHRE) [13].

The American Society for Reproductive Medicine (ASRM) has discussed the issue of surrogacy and published their recommendations recently [14] about the screening, evaluation, psychoeducational, and legal counseling of gestational carriers and intended parents.

In 2015, the European Parliament condemned this reproductive practice, considering that surrogacy constitutes an offense against women's dignity and promotes the instrumentalization of the surrogate's body and of her reproductive functions by treating her as an object of trade and making her vulnerable to abuse and exploitation.

In 2021 India introduced a new bill that prohibits commercial surrogacy, but allows altruistic surrogacy. Altruistic surrogacy involves no monetary compensation to the surrogate mother other than the medical expenses and insurance coverage during the pregnancy. The bill stipulates that LGBTQ+ families, single parents, unmarried couples, foreign citizens, and people outside the age groups of 23–50 for females and 26–55 for men are not permitted to seek surrogacy [15].

Responsibilities of Physicians to Pregnant Women Participating in Surrogacy and to the Intended Parents

When a woman participating in surrogacy seeks medical care for an established pregnancy, the obstetrician should explore, with the woman, her understanding of her contract with the intended parents and any provisions therein that may affect her care.

If the physician believes that the provisions of the contract may conflict with his or her professional judgment, the physician may refuse to accept the patient under those terms.

Once accepted as a patient, she should be cared for as any other obstetric patient, regardless of the method of conception, or else referred to an obstetrician who will provide that care. Even if she has already undergone screening by an agency, a physician–patient relationship exists between her and the obstetrician. The obstetrician has the attendant obligations resulting from this relationship.

Additional recommendations regarding the provision of obstetric services in this setting are as follows

- The obstetrician's professional obligation is to support the well-being of the pregnant woman and her fetus; to support the pregnant woman's goals for the pregnancy; and to provide appropriate care regardless of the patient's plans to keep or relinquish the future child. If a physician's discomforts with the surrogacy arrangements are seen to interfere with that obligation, the patient should be referred to another obstetrician.
- The pregnant woman should be the sole source of consent regarding clinical intervention and management of the pregnancy, labor, and delivery.
- Agreements the surrogate mother has made with the intended parents regarding her care or behavior during pregnancy and delivery should not affect the physician's care of the patient. The obstetrician must make recommendations that are in the best interests of the pregnant woman and her fetus, regardless of prior agreements between her and the intended parents.

- Confidentiality between the physician and the pregnant patient should be maintained. The intended parents may have access to the patient's medical information only with the pregnant woman's explicit consent.
- Obstetrician–gynecologists are encouraged to assist in the development of hospital policies to address labor, delivery, postpartum, and neonatal care in situations in which surrogacy arrangements exist.

Responsibilities of Infertility Specialists and Reproductive Endocrinologists to Intended Parents and Surrogate Mother

In providing medical services related to surrogate motherhood arrangements, infertility specialists and reproductive endocrinologists should follow the following recommendations

- A physician, who performs artificial insemination or in vitro fertilization as a part of surrogacy services, will necessarily be involved with both the intended parents and the surrogate mother. However, the intended parents and the surrogate mother should have both independent counseling and independent legal representation, and the surrogate mother should obtain obstetric care from a physician who is not involved with the intended parents.
- A physician who provides examinations and performs procedures for an agency that arranges surrogacy contracts should be aware of the policies of the agency and should decline involvement with any agency whose policies are not consistent with the ethical recommendations of Committee Opinion and those of other professional organizations related to reproductive medicine.
- Specialists in infertility and reproductive endocrinology are encouraged to participate in research that is intended to provide data on the outcomes of surrogacy arrangements [16].

The FIGO Standing Committee on Ethical Aspects of Human Reproduction discussed aspects of surrogate motherhood and made the following statement

1. The committee has strong reservations at the present time about the practice of surrogacy.
2. The committee was concerned that surrogacy, generally, might violate certain family values.
3. Surrogacy can be applied only in cases of very limited special indications (majority opinion).
4. The committee emphasized that special attention has to be paid to the ethical principle of protection of the surrogate mother who could be exploited because of her socioeconomic status.
5. The autonomy of the surrogate mother should be respected and the surrogate arrangement should not be commercial (organized by agencies).
6. Surrogacy, if conducted by individual physicians, should be approved by an ethical committee and practiced strictly under medical supervision.
7. When the practice is performed, it should take into consideration the laws of the country concerned, and participants should be fully informed of the legal position.

Medical Outcomes in Surrogacy

When we want to assess the complications involved in surrogacy, we must address the complications for both the intended parents and the gestational carrier.

The complications for the intended parents are only for the female partner and are associated with the IVF process including the risk of ovarian stimulation and oocyte retrieval.

The complications for the gestational carrier relate to both pregnancy and childbirth (even for a previously healthy woman) and may result in morbidity and mortality.

In a study [17], two out of ten gestational carriers required hysterectomies and blood transfusions (for placenta accreta and uterine rupture). Notwithstanding the above, most case series report no increase in adverse events related to surrogate pregnancy [18].

Multifetal gestation is more frequent among gestational carrier pregnancies than non-surrogacy IVF pregnancies. For example, the percentage of multifetal gestations out of all surrogate births in Israel between the years 1996 and 2017 ranged from 13 to 50% and averaged 23%. This fact is largely attributed to the transfer of multiple embryos, mainly due to the significant costs for the intended parents (including payments to the woman who is a gestational carrier, reimbursements for medical care, IVF costs, as well as agency and attorney fees) [19].

Medical complications more common in women with multifetal gestations include gestational diabetes mellitus, hypertension, anemia, hemorrhage, cesarean delivery, and postpartum hemorrhage [20]. Multiple gestation pregnancies are also associated with an increased risk for preterm birth and neonatal demise primarily because of the complications of prematurity. Because of these risks, the European Society of Human Reproduction and Embryology (ESHRE) and the American Society for Reproductive Medicine (ASRM) recommend that a single embryo is transferred [21, 22]. The ASRM, for example, recommends that "special consideration should be given to transferring a single embryo in an effort to limit the risks of multiple pregnancy for the carrier," but also notes that "after appropriate counseling and agreement by all parties, additional embryos may be transferred… in an effort to improve the probability of pregnancy" [23].

Some authors report lower rates of preeclampsia, low birth weight, and placental abruption in pregnancies achieved through gestational surrogacy compared with conventional IVF [17, 23], implying a protective role of a healthy carrier.

On the other hand, there are studies that compared commissioned gestational carrier cycles vs spontaneous cycles by the same gestational carrier and found an increase in adverse events, including gestational diabetes, hypertension, and placenta previa, and an increase in the likelihood of adverse perinatal outcomes in the commissioned cycles [24, 25]. This suggests that the process of ART may have adverse effects on the pregnancy.

As stated above, in the IVF process in young women, a single fetus is usually returned in order to avoid complications associated with multiple pregnancies for both the woman and the newborn. When we come to discuss this issue with regard to surrogacy, the recommendations should be identical due to the same medical considerations. Care should be taken not to prioritize economic and personal considerations in a way that would endanger the surrogate wife or the newborn.

Psychological Effects of Surrogacy

The Surrogate Mother

The motivation to participate in surrogacy can stem from both altruistic and personal/economic motives. The motive for surrogacy has a direct effect on the mental state of the surrogate mother. Moreover, a woman's motivation and desires are influenced by the culture in which she lives and by her socioeconomic status [19].

Thus, a significant difference can be seen between surrogates from developed countries and those from developing countries, such as India. While in Western countries, the main reason is altruistic, in India the main reason is financial. In a cohort study conducted in India, 15 carrier gestational women were interviewed. The motivation of these women was primarily financial. Moreover, they reported a severe stigma that forced them to leave the place where they lived until the end of the surrogacy process [26]. This situation led in 2018 to a change in the law in India that made surrogacy illegal.

Parker [27], in his study, reviewed the various motivations for surrogacy

1. A desire and need for financial benefit.
2. The enjoyment and desire to be pregnant.
3. A strong wish to produce a child as a gift to a parent in need.
4. A need to master unresolved guilt stemming from a previous voluntary abortion.

These observations were also noted in a different study [28]. The prospective surrogates appear to be women motivated by a mixture of personal and altruistic factors, without any notable pathology.

A comparison between pregnant surrogate and non-surrogate mothers indicated that surrogate mothers showed less attachment to the fetus, more positive attitudes toward certain dimensions of pregnancy, and less social support or more varied social support than non-surrogate mothers. Surrogate mothers had less support from their families than the non-surrogate mothers. The two groups were comparable in demographics and in pregnancy characteristics. The surrogate mothers had more positive attitudes toward their body image and to sex than did the comparison group. In general, surrogate mothers seemed to bond less and were therefore able to give up the baby more readily.

A pregnancy for a surrogate differs from a typical pregnancy. Some surrogates, at the beginning of the pregnancy, may feel satisfaction in having helped an infertile couple. As the pregnancy continues, scruples with regard to the growing embryo may evolve. If the woman is married, she may feel guilt toward her husband and children. After delivery, she may feel guilt toward the baby bond with the child and refuse to give it up for adoption.

The question arises as to whether true informed consent can actually be given by the surrogate or whether anyone can predict the emotions associated with relinquishing a child.

Screening of the prospective surrogate is vital, and in addition to adequate counseling, her emotional, mental, and physical suitability must be assessed by properly trained individuals. If the surrogate is the donor of an ovum, she must be genetically screened as well. It is necessary to screen her in order to ensure that she is not likely to engage in any activities during the pregnancy that could potentially harm the fetus.

A systematic review study has shown that surrogates suffer less from postpartum depression [28]. Moreover, Jadva et al. reported [29] that surrogate mothers do not appear to experience psychological problems as a result of the surrogacy arrangement. It showed that surrogate mothers did experience some problems immediately after handing over the child, but that these were not severe, tended to be short-lived and to dissipate with time. In a follow-up study by Jadva [30], which followed 20 of these surrogates 10 years later, they reported experiencing no long-term psychological problems; instead, many showed high levels of self-esteem.

The Child

The child may be confused as to who are his parents. He may have several prospective parents: the surrogate mother, her husband, and the infertile couple. This confusion may be further enhanced if the surrogate is also the mother of children. If the surrogate is a friend or relative who maintains contact with the child, it is unclear what effect this connection with two mothers will have on the child's development and identity.

It seems less likely that the use of a surrogate gestational mother would present self-identity problems than would the use of a donor egg or sperm, because the child would be reared by his or her genetic parents.

A child conceived through surrogate motherhood may be born into a much healthier climate than a child whose birth was unplanned. For this reason, some of the risks caused by confused genealogy may outweigh the possible benefits to the child of having parents who want him or her.

The Commissioning Couple

The commissioning couple may be distressed by the following problems: Is the contract which was signed with the surrogate mother legally valid? Will she give up the child for adoption? Will the child be healthy? The couple may be harmed by the surrogate's decision to keep the child, or, if she is a relative or friend, her continued involvement with the couple may cause tension in their marital relationship. Surrogacy could produce tension within the marriage because, although the male partner may be the genetic father, the female partner is usually not the genetic mother.

The Surrogate Woman's Husband and Children

The husband of the surrogate mother is the legal father until legal adoption has been processed. He may be affected mentally, physically, or legally. The same pertains to the surrogate's children.

A comparison between pregnant surrogate and non-surrogate mothers did not reveal any significant difference in the effect that the pregnancy has on the spouse. The marital relationship does not appear to be affected by the surrogacy. Likewise, according to the survey, a problematic marriage does not motivate a woman toward surrogacy [31].

A study that examined the effect of surrogacy on the surrogate's biological children showed that the children were found to have high levels of psychological well-being and reported close family relationships [32]. Moreover, the majority of the children felt positive about their mother's involvement in surrogacy.

The Social Aspects of Surrogacy

It might be argued that surrogacy adversely affects the way in which our society treats women and children, demeaning the place of both in society to the extent that they may be regarded as commodities. On the other hand, altruistic surrogacy may be seen as a supreme example of charity which, in turn, could provide the stimulus for society as a whole to become more charitable.

Surrogacy may threaten the status of marriage in the community, because of the analogy to adultery and the separation of love-making and reproduction, the latter becoming a mere technological matter divorced from the psychological and emotional aspects of sexual intercourse.

Legalizing Surrogacy

The legal status of surrogacy varies greatly from one country to another, with two main types of regulation

1. Surrogacy is regulated by legislation. Legalization of gestational surrogacy aims to defend the surrogate's interests as well as those of the intended parents and the baby born after the surrogacy. Most countries do not have legislation.
2. Surrogacy is practiced by guidelines based on the types of contractual arrangements between parties and includes either a commercial or altruistic transaction depending on whether the surrogate receives a financial reward for her pregnancy or not.

Legalization of gestational surrogacy aims to defend the surrogate's interests as well as those of the intended parents and the baby born as a result of surrogacy.

Legalizing Surrogacy in Israel [33–36]

In 1991 the Ministers of Health and of Justice in Israel nominated a public committee, the Aloni Public Committee (this article's co-author, J. G. Schenker, as member of the committee, prepared the minority report). The committee's tasks were to inspect the social, ethical, religious, and legal aspects of in vitro fertilization (IVF). It was commissioned to prepare a proposal for legislation, with special attention to the question of the surrogate mother.

Reporting in 1994, its main recommendation was that surrogate motherhood through IVF should be permitted but had to be regulated, primarily to obtain a priori approval from a statutory body. This liberal recommendation reflected the belief of the majority of the committee that principles of "autonomy" and "privacy" require minimum state interference in human reproduction.

According to the committee's recommendations, the Israeli parliament, the Knesset, passed the law concerning surrogacy in March 1996, *Embryo Carrying Agreements Law (Law 5756, 1996)*.

The new surrogacy law in Israel is the first state law concerning infertility, as previous legislation was based on current legal regulations established by the Ministry of Health. The new law resulted in a compromise between the restrictions of the Orthodox regulations and the liberal views of secular Israelis. At that time, it was the only national law apart from that of the United Kingdom allowing state-controlled surrogacy.

Under the new law in Israel, every single case of surrogacy must be authorized by a special committee. The State-Appointed Permission Authorization Committee is a multidisciplinary committee, nominated by the Minister of Health, and comprises seven members:

1. Two physicians qualified in obstetrics and gynecology.
2. A physician qualified in internal medicine.
3. A clinical psychologist.
4. A social worker.
5. A lawyer, representing the public interest.
6. A clergyman, representing the religion of the involved parties.

The decisions of the committee must be accepted by the majority of its members and must be made in the presence of at least five of the members, including the chairman.

Guidelines Set by the Committee for Surrogacy

The guidelines include the following

1. Full surrogacy is permitted only when the gametes are provided by both parties of the commissioning couple (CC), who are married according to the law of the country. In special cases, the committee can authorize surrogacy with ovum donation.
2. Sperm donation is not allowed, since according to Jewish law, the child would be classed as "illegitimate."

3. The parties in the agreement are adult Israeli citizens (a clause designed to prevent the abuse of women from underdeveloped countries or illegal commercialization of the procedure).
4. The surrogate mother should be single, widow, or divorced; otherwise, the child is "illegitimate" according to Jewish law. In very extreme cases, the committee can authorize an agreement involving a married woman.
5. The surrogate mother is anonymous and not a relative of one of the parents. (Relatives such as mother daughter, granddaughter, sister, aunt or cousin are forbidden. Adopted relatives are permitted.) This avoids any potential pressure on relatives to become surrogate mothers with subsequent complications within the family.
6. Since the Israeli population is made up of multi-ethnic and multi-religious groups including Jews, Moslems, Christians, and others, the attitude of the various religions was considered. According to Jewish law, the religion of the child is determined by the religion of the mother. Therefore, the surrogate mother and the CC should be of the same religion, although if all the parties are not Jewish, the committee may allow an interreligious agreement following consultation with the clergymen members of the committee.

However, since Moslems and Christians do not allow surrogacy, they are unlikely to approve such an agreement.

Preconditions for Approving Surrogacy by the Approving Committee

1. A medical report should be presented, stating that the mother of the CC is either unable to become pregnant or carry a pregnancy to term or that a pregnancy could be a major risk to her health.
2. A medical opinion must approve the suitability of both parties involved in the process.
3. The surrogate mother should receive a general and gynecological check-up, to rule out diseases that could be aggravated by pregnancy and delivery. A history of drug use, alcohol abuse, or medications that could affect pregnancy should be obtained.
4. The gynecological history should exclude medical conditions that may cause either early or late abortions, early deliveries, or any other complication of the pregnancy.
5. Tests for transmissible diseases, including human immunodeficiency virus (HIV), hepatitis B and C, and VDRL should be performed, together with blood group analysis in order to prevent rhesus incompatibility. Ultrasound examination of the pelvis must exclude major malformations of the uterus and of other pathologies of the uterus or the cervix.
6. A psychological assessment of the parties involved must be provided, followed by a statement by a psychologist or a social worker that the CC has received suitable professional guidance, in which other possibilities for parenthood were discussed.

7. The procedure is to be performed in a certified in vitro fertilization-embryo transfer department.
8. If the parties were selected through a paid mediator, the agreement with the mediator, including his name, should be presented to the committee.

Expenses

One of the main objects of the Approving Committee is to prevent illegal commercialization of the procedure. The committee supervises the agreement of expenses and can recommend monthly payments to the surrogate mother to cover actual expenses including medical coverage, insurance, legal consultation, loss of time and income, or any other reasonable compensation.

The Legal Status of the Newborn

Only following the court's approval, will the CC become the child's sole parents for all matters. In extreme cases, where the child has a malformation at birth and becomes HIV positive during pregnancy, or in any other circumstances where the CC withdraws from the agreement, the surrogate mother becomes the legal guardian of the newborn. In cases where the surrogate mother refuses to raise the child, it will be transferred to the state welfare authorities.

Surrogate Mother Withdrawal from the Agreement

The court will approve the withdrawal of the surrogate mother from the agreement only in cases where the social worker's report provides convincing evidence of a change in circumstances that provides justification and where the well-being of the child would not be compromised. After the adoption act has been finalized, the court cannot authorize withdrawal from the agreement. In cases wherein the court approves the withdrawal, it should nominate the surrogate mother as the only legal mother and guardian of the child. The court can also rule upon the relationship between the child and one or both members of the CC.

In event that the court has granted custody to the surrogate mother, it can order the repayment of expenses to the parties.

Legal Rights of the Surrogate Mother

The rights of the surrogate mother: The law should not contradict the rights of the surrogate mother to medical treatments or prevent her from either procuring or performing a medical procedure of her own free will, including interruption of pregnancy under the terms set by the law. Neither the CC nor any other body have the right to control the surrogate's way of life during pregnancy, including nutrition, drinking habits, sexual behavior, or use of drugs. The CC cannot intervene in the prenatal care received by the surrogate mother, nor can it force her to undergo invasive and noninvasive perinatal procedures, such as amniocentesis, against her will.

Enforcement of the law: According to this law, establishment of a surrogacy agreement without the Authorization of the Permitting Committee is a criminal offence, carrying a prison sentence of 1 year.

The right to privacy: Publication of details from the committee's discussions—which include information that could identify any of the parties involved—is forbidden and carries a prison sentence of 1 year.

Illegal financing: A party that offers, gives, or demands money or benefits for participation in such an agreement without the approval of the committee is committing a criminal offence.

Legal adoption: Deliverance or acceptance of a child without the presence of the social worker or without a court order carries a prison sentence of 1 year.

While adopting the commission's recommendation for approval of agreements, Law 5756, 1996 ("the Law") diverges from the views of the majority in relation to a number of important issues.

In particular, while the commission envisaged that surrogacy would be largely "altruistic," the effect of the law is that surrogacy will invariably occur on a commercial basis. For example, the law prohibits relatives of the intended parents to serve as surrogate mothers. Furthermore, while the commission envisaged that the surrogate mother would be reimbursed for financial expenses and losses incurred during the process, though not receive any actual payments, the law allows payments to compensate her for her time and suffering.

Other fundamental differences are that the law does not allow partial surrogacy and that, while the majority of the commission recommended that the handing over of the child to the intended parents should be sufficient to determine her/his status as their child, the law requires a parentage order to be made by a court.

In accordance with the law, regulations were published dealing with the technical aspects of the implementation of the law and included standard forms for completion by the parties before and after the birth.

The Authorization Committee has issued Guidelines to Applicants, detailing the documents which have to be submitted, the forms which have to be completed and conditions which have to be included in the contract.

The Legal Aspects of Surrogate Practice

The implementation of surrogate motherhood in an IVF program gives rise to several legal problems [2].

The surrogate mother may refuse to give up the child, or the genetic parents may refuse to take him in cases of physical or mental defects. Do genetic parents have the right to compel the surrogate mother to undergo amniocentesis or abortion in cases of abnormalities in the fetus? Does the surrogate mother have the right to undergo abortion if she changes her mind? Do the genetic parents have the right to control the surrogate's life-style during the pregnancy (nutrition, drinking habits, sexual life, drug abuse, etc.)? Such questions may be solved in contracts between the surrogate mother and the commissioning couple, but the question remains as to the legality of such a contract.

Another question is: Who is the legal mother? In partial surrogacy, there is no dispute, since the legal mother is the surrogate. In full surrogacy, however, the woman who bears the child is the surrogate, while the biological mother is the commissioning parent. According to international law, the woman who bears the child is his legal mother, regardless of who donated the oocyte. The result is that in either situation, the offspring must be legally adopted by the commissioning mother.

According to the Israeli law, surrogacy as a route to parenthood is currently open to heterosexual couples and to single women who are connected genetically to the baby.

In February 2020, the High Court of Justice challenged a controversial law that prevents single men and gay couples from using surrogacy to have children and gave the Knesset a year to pass a new law. This would allow Israeli male homosexual couples, single Israeli men, and transgender individuals to arrange surrogate pregnancies in the country. In January 2022, the Israeli law regulating surrogacy was updated so that it allows same-sex couples to go through a surrogacy process.

Israeli Ministry of Health Data 1996–2017

Graph 1 shows the number of applications submitted to the Board for Approval of Fetal Carriage Agreement in Israel and the numbers of births between the years 1996 and 2017.

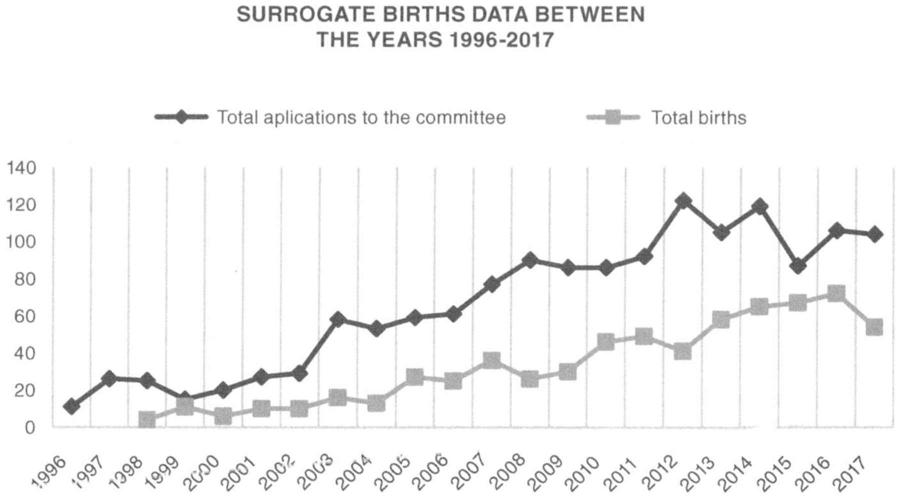

In total there were 1458 applications and a total of 666 births, resulting in the delivery of 823 children.

The percentage of multifetal gestations out of all surrogate births between the years 1996 and 2017 ranged from 13 to 50% and averaged 23%. Of these pregnancies, there were even two triplet pregnancies.

Regarding the age of the surrogates, 7% were between the ages of 22 and 25, 23% were between the ages of 26 and 30, 47% were between the ages of 31 and 35, and 23% were between the ages of 36 and 39. 65% were married at the time of the surrogacy.

Regarding the education of the surrogates, 32% held an academic degree of which 9% held a master's degree or doctorate, 59% completed 10–13 years of study, and 8% held a professional diploma.

Religious Aspects of Surrogacy

Judaism

The Israeli legislation on surrogacy is partly based on the Jewish law, Halakha. There are three basic principles that, with certain restrictions, favor the acceptability of the practice of surrogacy: firstly, the commandment "Be fruitful and multiply" and secondly, the mitzvah of benevolence (G'miluth hasadim), which originates in the verse "Love thy neighbor as thyself" (Leviticus 19:8). In cases of personal distress (material, mental, or both), a Jew is duty-bound to help him fulfill this commandment. A childless couple will fall within this category in which a clear obligation exists to assist them in every permissible way, as long as no one else is thereby harmed. Thirdly, domestic harmony and the integrity of the family are extremely important in Jewish law.

The Jewish religion does not forbid the practice of surrogacy. According to Judaism, in cases of partial surrogacy, the father of the child is the sperm donor, and therefore he is the biological and social father.

From the religious point of view, in cases of full surrogacy, which was created by the donation of oocyte and sperm by the commissioning couple and transferred to the surrogate, the child will belong to the donor of the sperm and to the mother who gave birth.

On the other hand, Jewish Halakha also presents some problems for surrogacy, which are only partially resolved. The principal problems are illegitimacy (mamzerut) and the risk of marriage between siblings in the future, which would amount to incest.

Christianity

The practice of surrogacy is unacceptable to the Christian churches—Roman Catholic, Greek Orthodox, Protestant, Anglican, and most other congregations. The objection is based on the belief that surrogacy is contrary to the sanctity of marriage and to the dignity of human procreation.

Islam

Surrogacy is unacceptable in Islam on the premise that pregnancy should be a fruit of a legitimate marriage. If surrogacy were to be practiced by Moslems, the child delivered would belong to the woman who carried it and gave birth to it, since the Koran declares that "Our mothers are those women who provide the womb and give birth." The proposal that surrogacy could be practiced among multiple wives of a Moslem husband was recently rejected. It should be mentioned that adoption is forbidden by Islam.

Hinduism

Hinduism is a diverse body of religion, philosophy, and cultural practice predominant in India, characterized by a belief in reincarnation and a supreme being of many forms and natures. Hindu believers are governed by the three doctrines of dharma, or universal law; karma, or the cumulative effects of personal actions; and samsara, or the liberation from which is the first goal of life.

Hinduism has no single book, such as the Bible, that serves as the source of its doctrine, but it has many writings, all of which have contributed to its fundamental beliefs.

There is a huge stigma attached to being infertile in Indian society, especially for the woman. Regarding fertility, the emphasis on reproduction is not just on having children but on having male offspring to continue the family line and perform religious rituals for the salvation of departed souls.

ARTs are acceptable in Hinduism because there is no single authority to accept or reject it on behalf of the faith including oocyte and embryo donation, surrogacy, and sex preselection. India became a leading country for reproductive tourism, especially for surrogacy. Until recent change in legislation (?), the reasons for the surrogacy boom in India were the relatively low costs and easy availability of women for surrogacy, especially those from socioeconomically disadvantaged backgrounds.

Buddhism

Buddhism, one of the major religions of the world, founded in India about 500 BC by the Buddha. At various times, Buddhism has been a dominant religious, cultural, and social force in most of Asia, especially in China, India, Japan, Korea, Tibet, and Vietnam. In each area, Buddhism has combined with elements of other religions such as Hinduism and Shinto.

All Buddhists have faith in the Buddha; his teaching, called the dharma; and the religious community he founded, called the sangha. The basis of what Buddha preached in the dharma is that existence is a continuing cycle of death and rebirth.

Buddhism has never been organized around a central authority; therefore, Buddhists of all types in various countries are individualistic, and even their scriptures are not rigid. There is no central Buddhist authority to pronounce religious positions.

Marriage within Buddhism does not have the high priority that it has in monotheistic religions. According to Buddhism, the three factors necessary for the rebirth of a human being are the female ovum, the male sperm, and the karma. This karma energy is sent forth by the dying individual at the moment of his or her death.

Any technology that is used to achieve conception is morally acceptable including surrogacy, and treatment may be given to both unmarried and married women according to governmental limitation where it exists.

Conclusions

Although very few infertile couples need to resort to IVF surrogacy to help them have a child, the subject has nevertheless provoked a lot of discussion over the past four decades. There is an ongoing debate in ethical literature regarding commercial surrogacy. Those in favor of surrogacy regard it as an ethical concept of free choice and personal autonomy. The opponents argue that the moral nature of child-bearing and the parent-child relationship is negatively affected by commercialized surrogacy agreements.

There is a fundamental distinction between paternity and maternity in surrogacy. While paternity is based on the genetic and only on the genetic function, maternity normally has two aspects, the genetic one, of providing the oocyte and the physiological function: one of gestation and parturition.

Surrogacy practice may affect individual liberty, equality, and the family. On the one hand, surrogacy appears to reinforce the traditional family by allowing infertile married couples to create biologically related children. On the other hand, surrogacy possesses the potential to radically destabilize and disrupt the traditional concept of the family.

While surrogacy is represented as a last-resort medical "solution" to the problem of infertility, the varying international responses to its regulation indicate that it is more often perceived as a social "problem."

Each society (country) formulates its own most appropriate solution. The Israeli legislative and administrative framework regulating surrogacy arrangements is designed to protect all the parties involved. The Israeli law (1996) gives priority to the interests of the child in every aspect of a surrogacy scheme. The child, whose involvement is involuntary, is the most vulnerable of all the parties, and thus his/her interests must be the main concern at all stages of the process.

As a summary of all the aspects reviewed in this chapter, it can be seen that in the vast majority of cases, there is a positive effect of the surrogacy process on all concerned—the biological parents, the surrogate and her environment, and the newborn who would not have been born without this process. It should be emphasized that this process needs to be supervised so that its goals are indeed maintained and there will be no harm to all concerned.

References

1. British Medical Association. Council of the British Medical Association Annual Report 1989–1990. Appendix V. London: British Medical Association; 1990.
2. Schenker JG, Frenkel DA. Medico-legal aspects of in-vitro fertilization and embryo transfer practice. Obstet Gynecol Surv. 1987;41:405–13.
3. Schenker JG. The surrogate mother. Harefuah. 1992;123:330–3.
4. Old Testament . Book of Genesis. 16:1–6.
5. Old Testament. Book of Genesis. 30:1–4.
6. Keane N, Bred D. The surrogate mother. New York: Everest House; 1981.
7. Corea G. The mother machine. New York: Harper and Row; 1985.
8. Utian WH, Sheean L, Goldfarb JM, Kiwi R. Successful pregnancy after in vitro fertilization and embryo transfer from an infertile woman to a surrogate. N Engl J Med. 1985;313:1351–2.
9. British Medical Association. Annual report of council. Appendix V: surrogacy report. London: British Medical Association; 1990. p. 39–48.
10. Warnock M. Report of the committee inquiring into human fertilization and embryology. London: Her Majesty's Stationery Office; 1984.
11. FIGO Standing Committee for the Study of Ethical Aspects of Human Reproduction. Recommendation on surrogate motherhood. Int Gynecol Obstet. 1989;28:299.
12. Ethics Committee of the American Fertility Society. Surrogate gestational mothers: women who gestate a genetically unrelated embryo. Ethical Considerations of the New Reproductive Technologies. Fertil Steril. 1986;46(Suppl. l):1s–94s.
13. ESHRE Task Force on Ethics and Law. ESHRE task force on ethics and law 10: surrogacy. Hum Reprod. 2005;20:2705–7.
14. ASRM Recommendations to Practice. Using gestational carries. Fertil Steril. 2022;118:70.
15. The Surrogacy (Regulation) Act, 2021 (No. 47 of 2021) [25th December, 2021].
16. Yau A, Friedlander RL, Petrini A, Holt MC, White DE 2nd, Shin J, Kalantry S, Spandorfer S. Medical and mental health implications of gestational surrogacy. Am J Obstet Gynecol. 2021;225(3):264–9.
17. Swanson K, Debbink M, Letourneau JM, Kuppermann M, Einerson BD. Association of multifetal gestation with obstetric and neonatal outcomes in gestational carrier pregnancies. J Assist Reprod Genet. 2021;38(3):661–7.
18. Committee on Practice Bulletins Obstetrics, Society for Maternal Fetal Medicine. Practice Bulletin No. 169: Multifetal gestations: twin, triplet, and higher-order multifetal pregnancies. Obstet Gynecol. 2016;128:e131–46.
19. Shenfield F, Pennings G, Cohen J, et al. ESHRE task force on ethics and law 10: surrogacy. Hum Reprod. 2005;20:2705–7.
20. Practice Committee of the American Society for Reproductive Medicine, Practice Committee of the Society for Assisted Reproductive Technology. Recommendations for practices utilizing gestational carriers: a committee opinion. Fertil Steril. 2017;107(2):e3–10.
21. Parkinson J, Tran C, Tan T, Nelson J, Batzofin J, Serafini P. Perinatal outcome after in-vitro fertilization-surrogacy. Hum Reprod. 1999;14:671–6.
22. Woo I, Hindoyan R, Landay M, et al. Perinatal outcomes after natural conception versus in vitro fertilization (IVF) in gestational surrogates: a model to evaluate IVF treatment versus maternal effects. Fertil Steril. 2017;108:993–8.
23. Pavlovic Z, Hammer KC, Raff M, et al. Comparison of perinatal outcomes between spontaneous vs. commissioned cycles in gestational carriers for single and same-sex male intended parents. J Assist Reprod Genet. 2020;37:953–62.
24. Karandikar S, Gezinski LB, Carter JR, Kaloga M. Economic necessity or noble cause? A qualitative study exploring motivations for gestational surrogacy in Gujarat, India. Affilia. 2014;29:224–36.
25. Parker PL. Motivation of surrogate mothers: initial findings. Am Psychiatry. 1983;140:117–8.

26. Soderstrom-Anttila V, Wennerholm U-B, Loft A, Pinborg A, Aittomaki K, Romundstad LB, et al. Surrogacy: outcomes for surrogate mothers, children and the resulting families—a systematic review. Hum Reprod Update. 2016;22:260–76.
27. Jadva V, Murray C, Lycett E, MacCallum M, Golombok S. Surrogacy: the experiences of surrogate mothers. Hum Reprod. 2003;18:2196–204.
28. Jadva V, Imrie S, Golombok S. Surrogate mothers 10 years on: a longitudinal study of psychological well-being and relationships with the parents and child. Hum Reprod. 2015;30:373–9.
29. Fischer S, Gillman I. Surrogate motherhood: attachment, attitudes and social support. Psychiatry. 1991;54:13–20.
30. Jadva V, Imrie S. Children of surrogate mothers: psychological well-being, family relationships and experiences of surrogacy. Hum Reprod. 2014;29:90–6.
31. Schenker JG. Assisted reproduction: formation of laws and regulations. In: Boutaleb Y, Gzouli A, editors. New conception in reproduction. Carnforth: Parthenon Publishing; 1991. p. 149–58.
32. The Aloni Commission Report, 1994. (Hebrew)
33. Schenker JG, Eisenberg V. Surrogate pregnancies. Prenat Neonatal Med. 1996;1:29–37.
34. Benshushan A, Schenker JG. Legitimizing surrogacy in Israel. Hum Rep. 1997;12:1832–5.
35. Israeli Ministry of Health: Surrogate Committee Report 1996–2017. (Hebrew).
36. Schenker JG. Surrogate motherhood: international perspectives. In: Cook R, Sclater S, editors. Legitimizing surrogacy: religious perspectives. Oxford: Hart Publishing; 2003. p. 243–61.

Human Reproductive Cloning

8

Giuseppe Benagiano and Paola Bianchi

Introduction

This chapter aims at summarizing salient issues in human reproductive cloning (HRC), defined as "*The use of technologies, including somatic cell nuclear transfer* (SCNT), *to create offspring with the shared genomic material of the original person*" [1]. In the course of this exposé, mention will also be made of human therapeutic cloning (HTC), defined as a method that "*uses these same experimental techniques for therapies other than reproduction* (*such as research, production of embryonic stem cell lines, or creation of solid organs for transplant*)" [1].

In order to position reproductive cloning within the realm of biology, it is opportune to stress that in the animal kingdom, reproductive processes are so diversified to include any conceivable mechanism: reproduction can be bisexual with internal (e.g., mammals) or external (e.g., invertebrates, amphibians, fish) copulation; in addition, there is the so-called *sequential hermaphroditism*, when in a species sex can be interchangeable (e.g., bearded dragons, red frogs, clownfish), *true hermaphroditism* (e.g., worms, moss animals, snails), and even *parthenogenesis* (e.g., some lizards and crustaceans).

This means that, at least in principle, reproduction by cloning may be considered as one of the many natural options. However, in evolutionary terms, each and every species developed over millennia its particular form of reproduction to better suit its needs. For mammals in general and the genus *Homo* in particular, evolution produced the arguably only mechanism capable of increasing diversity and improving

G. Benagiano
Geneva Foundation for Medical Education and Research, Geneva, Switzerland

P. Bianchi (✉)
Department of Medico-Surgical Sciences and Translational Medicine, Sant'Andrea Hospital, Sapienza University of Rome, Rome, Italy
e-mail: paola.bianchi@uniroma1.it

J. G. Schenker et al. (eds.), *Hot Topics in Human Reproduction*, Reproductive Medicine for Clinicians 3, https://doi.org/10.1007/978-3-031-24903-7_8

the species. This is why it can be affirmed that, over and above the ethical reasons that will be discussed in detail, among humans, reproduction by cloning is directly against the evolutionary pathway set millions of years ago.

It can be argued that, with the advent of assisted reproduction technology (ART), a number of mammals, as well as humans, moved away from the evolutionary path set for our species. This, however, is only partially true, since the scope of ART is bisexual reproduction, albeit achieved through external fertilization and without copulation.

It has been said that, having accepted external fertilization, the next step, reproductive cloning when there are no sperm or eggs, may become acceptable, especially when a couple is opposed to sperm, or oocyte, donation or adoption.

More recently, work has been carried out on a technique, *in vitro gametogenesis* (IVG), that could possibly rectify germ cell aplasia (e.g., non-obstructive azoospermia and oocyte maturation failure syndrome). Today, primordial germ cells can be derived from pluripotent stem cells, although further progression to post-meiotic germ cells usually requires a gonadal niche and signals from gonadal somatic cells [2]. It has been argued that, "*if safety is the main reason for not allowing reproductive cloning, one might expect a similar conclusion for the reproductive application of IVG, since both technologies hold considerable and comparable risks*" [3]. This may be true, but, as the authors concede, risk is not the sole or even the main reason why cloning is being condemned. In fact, proponents of HRC call it *reproductive regeneration*, a term that, ironically, perfectly describes why there is such a widespread rejection of the technique: the fact that it negates the very essence of reproduction, *generation*, not *re-generation*.

As mentioned by the Ethics Committee of the American Society for Reproductive Medicine (ASRM), the prospect of using HRC has produced an intense debate involving "*lawmakers, academicians, ethicists, religious leaders, international and national agencies, professional societies, and others*" [1]. In the end, decisions will be based on two fundamental issues: on the one hand the safety and efficacy of the procedure and on the other laws or governmental regulations based on the intensity and extent of ethical objections.

Technical Aspects

The word "clon" was first used in the nineteenth century in botany, with the final 'e' added in 1903. It referred to the asexual propagation of any plant, mostly by replanting cuttings. The word was subsequently extended to natural/asexual, molecular, cellular, and artificial reproduction [4]. The word originates from the ancient Greek word κλών, meaning "twig," and is today utilized in biology to identify a group of identical entities and, more specifically, an organism that is a genetic copy of another organism. The term is utilized to identify the "copy" of an entire organism, as well as "copies" of molecules (such as DNA) and cells.

As mentioned, in nature cloning occurs in those species that produce their offspring without combining male and female genetic material. During the twentieth century, biologists have attempted to artificially clone first amphibians and, with the advent of the twenty-first century, also mammals. The first mammalian born through cloning was the famous sheep Dolly in Scotland [5]. The absolute novelty consisted in the fact that her embryo was created using mature cells taken from an adult sheep mammary tissue. Although this achievement was hailed as a major revolution in reproduction, Dolly's premature illnesses led to the conclusion that she was afflicted by conditions typical of old age [6]. A genomic analysis of her DNA seemed to support the hypothesis of a premature aging due to telomere shortening [7]; however, in 2021, a concise, but accurate, summary of the situation [8] stressed that this finding contrasted with a number of investigations that generally found "*telomeres to 'rejuvenate' during nuclear reprogramming*." In addition, several studies have now concluded that "*cloned offspring which survive beyond the neonatal period are healthy, age normally, produce viable offspring and animal products safe for human consumption*" [9].

Yet, enough concerns remain, and, as of 2022, a ban is enforced on commercial farm-animal cloning within the EU and the UK, but not the USA and a number of other countries.

With regard to human cloning, as we will see, these concerns have led to an outright ban on HRC at the international level.

The technique that allows cloning in mammals has been coined *somatic cell nuclear transfer* (SCNT). It consists of transferring the nucleus from a donor cell into an oocyte or an early embryo from which the chromosomes have been removed; depending on the species, there are many variations in the details of the method [10]. Initially, researchers envisaged to use SCNT as a way to determine whether genes remain functional even after most of them have been switched off when a specific type of cell starts to carry out a specialized function. In this respect, the fact that the DNA of a fully differentiated cell could revert to an undifferentiated status and become capable to initiate the process of embryonic development would demonstrate that all the genes in differentiated cells retain their functional capacity, although only a few remain active. Therefore, in terms of developmental biology, the greatest outcome of the new technique has been the discovery that yet-unknown factors in the recipient oocyte can reprogram the nucleus to a very early developmental stage. The technique has now been refined, with a major advance consisting in the use as the recipient cell of an oocyte enucleated at meiotic metaphase II [11]. Obviously, the critical step is the activation of the enucleated oocyte following insertion of the new nucleus. In this respect, during the physiological fertilization process, activation is induced by the release of intracellular calcium in a series of pulses that follow the sperm entry. Artificially, activation can be obtained through exposure to ethanol, strontium, or pulses of alternating current [10].

In 2003, two cloned rhesus monkeys were born [12], but success was obtained only following transfer of nuclei from 4- to 8-cell stage embryos and none following transfer at a later stage. This failure was attributed to removal during the process of enucleation of key factors from the oocyte.

Finally, in 2011, an experiment was carried out to exchange the genome of a human oocyte with that of a somatic cell; the experiment aimed at producing pluripotent stem cells to be used for cell replacement in subjects with degenerative human diseases. In the original experiment, the development of human oocytes after

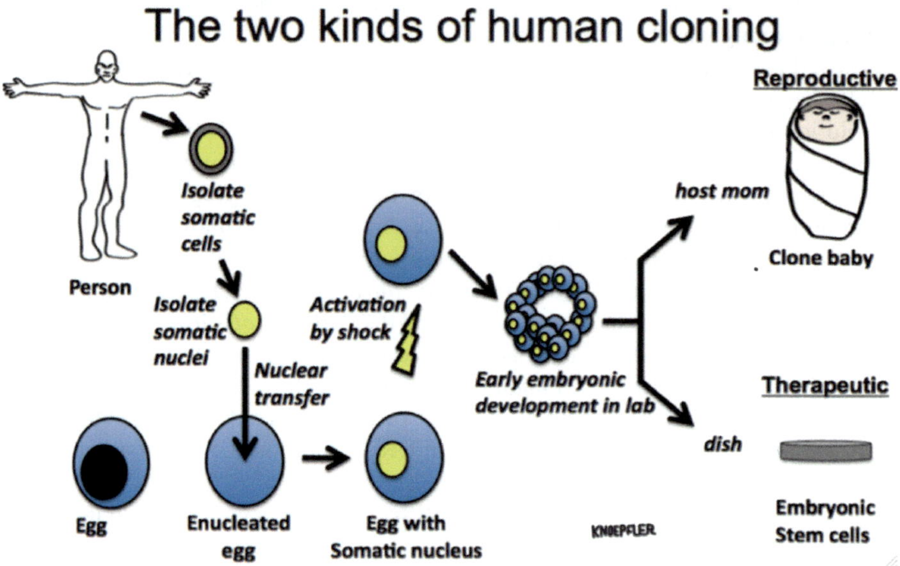

Fig. 8.1 Schematic representation of the two variant forms of human cloning: the first (***Reproductive cloning***) aimed at obtaining an individual with the same genetic and physical characteristics of the donor; the second (***Therapeutic cloning***) aimed at producing totipotent embryonic cell for research and therapeutic purposes (Reproduced from "The Niche" 2013, with permission). Schematic representation of the two variant forms of human cloning: the first (***Reproductive cloning***) aimed at obtaining an individual with the same genetic and physical characteristics of the donor; the second (***Therapeutic cloning***) aimed at producing totipotent embryonic cell for research and therapeutic purposes (Reproduced from "The Niche" 2013, with permission)

genome exchange arrested at late cleavage stages in association with transcriptional abnormalities [13].

If the blastocyst resulting from SCNT is transferred into the uterus of a female host and pregnancy progresses to term, the resulting individual will be a *clone*, since it will carry the same nuclear genetic material as the donor (Fig. 8.1). The expression "nuclear genetic material" of the adult somatic cell indicates that the offspring would not be an exact copy, because of the presence in the oocyte cytoplasm of a set of mitochondria, representing a prominent source of energy metabolism, but also containing a specific type of mitochondrial DNA that will later populate the cells of the offspring [14].

In sexual reproduction, clones are created when a fertilized egg splits to produce identical (monozygous) twins with identical genomes.

Attempts to Achieve Human Cloning

John Burdon Sanderson Haldane, a British scientist who was one of the founders of Neo-Darwinism, was the first to have thought of the possibility of human cloning and believed that it would 1 day be utilized to create super-human, super-talented individuals. He wrote: "*Assuming that cloning is possible, I expect that most clones would be made from people aged at least fifty, except for athletes and dancers, who would be cloned younger. They would be made from people who were held to have excelled in a socially acceptable accomplishment. ... Other clones would be the asexual progeny of people with very rare capacities, whose value was problematic, for example permanent dark adaptation, lack of the pain sense, and special capacities for visceral perception and control. Centenarians, if reasonably healthy, would generally be cloned, if this is possible; not that longevity is necessarily desirable, but that data on its desirability are needed*" [15].

In fact, in spite of this prediction, today there is an almost ubiquitous opposition to cloning for reproductive purposes coming from both the scientific community and the public at large. Haldane may have been aware that his words would 1 day be rejected, since he concluded: "*I'm not a biologist or a botanist, so I apologize if any of the above is ill-informed or incorrect, and I would be happy to be corrected, or to learn more*."

Because of the strong opposition, over the last decade, the few who are determined to proceed along the path leading to the birth of a cloned baby have been working in an atmosphere of mystery and secrecy not conducive to true scientific advances.

Since the turn of the millennium, there have been suggestions that HRC may represent a way to improve the human genetic endowment of mankind by cloning individuals of great achievements. Although these suggestions have generally never been taken seriously, some physicians have on occasion made clear that they were ready to carry out cloning [16], giving rise to a number of sensational reports on such attempts.

On 11 April 2002, Alison Abbott, in a short note in *Nature* [17], reported that European scientists had voiced skepticism about claims by the Italian gynecologist Severino Antinori that one of his patients was 2-month pregnant with a cloned human embryo. Ian Wilmut, senior member of the team that cloned Dolly the sheep, labeled the claim as "*either a misunderstanding, or deliberately misleading*." The following month, Antinori told the Italian magazine *Oggi* that three clones already existed: two boys and a girl who at the time were 9 years old and living in Eastern Europe. But, as usual, he provided no proof to confirm any of his claims! Then on 23 June 2002, the *Chicago Tribune* published that an "*informal consortium*," led by Antinori and the American andrologist Panayiotis Zavos, "*has currently 39 women in treatment*" to have a cloned baby and that 5 of these women were actually pregnant.

During December of 2002, the Geneva-based Raelian cult, which believes that humans were originally created by aliens, claimed that a baby girl named Eve had been born from an egg fertilized using a skin cell from her mother [18]. Allegedly, the cloning was carried out by a research outfit separate from the sect named *Clonaid*, and a few days later it was announced that a second clone, a girl, had been created and was born from a Dutch lesbian woman. However, Eve did not undergo genetic testing to compare her DNA to that of her mother's, the only way to prove or disprove the claim. As expected, no confirmation of these claims was ever published, and no DNA testing of the alleged second infant was ever carried out.

Unfortunately, the *tam-tam* of fake news continued, and, in a comment dated 22 April 2009, Andy Coghlan [19] stigmatized the "clone-mania," mentioning the claim by Zavos to have produced 14 cloned human embryos and transferred 11 of them into the wombs of women. He explained that, while none of the embryos survived this time, "*the cloned child is coming*," since "*there is absolutely no way that it will not happen*."

Once again, there is a lack of scientific information supporting these claims, and a search of *PubMed* in January 2022 found no publication whatsoever by "Zavos and Antinori." When searching for "Antinori S.", 282 results were obtained; 15 were by Severino Antinori (starting in 1991), all totally unrelated to human cloning. All others were from three homonymous researchers. As for "Zavos P. M.," a total of 92 results were obtained. Two articled dealt with HRC: in the first [20], he claimed to "*have never stated that we intended to create the first cloned embryo and the first human being for reproductive purposes by ignoring the public's concerns and the scientific critics. We also never intended to ignore the contradictory results that scientists in the field of animal cloning have obtained during the past years*." Zavos went on describing how his team "*created the first human cloned embryo*," as "*the end result of using nine microsurgically enucleated human donor oocytes and fusing them via electrical stimulation and activation with whole human granulosa cells from a patient desiring to have a child via SCNT*." He went on mentioning that "*the resulting cloned embryo was allowed to develop further in culture for 4 days post-SCNT and reached the 8–10-cell stage*." He concluded: "*its development was observed and recorded, and the embryo was cryopreserved for future molecular analysis and other observations*" (see Fig. 8.2). Not surprisingly, his promise: "*Full*

Prentice DA

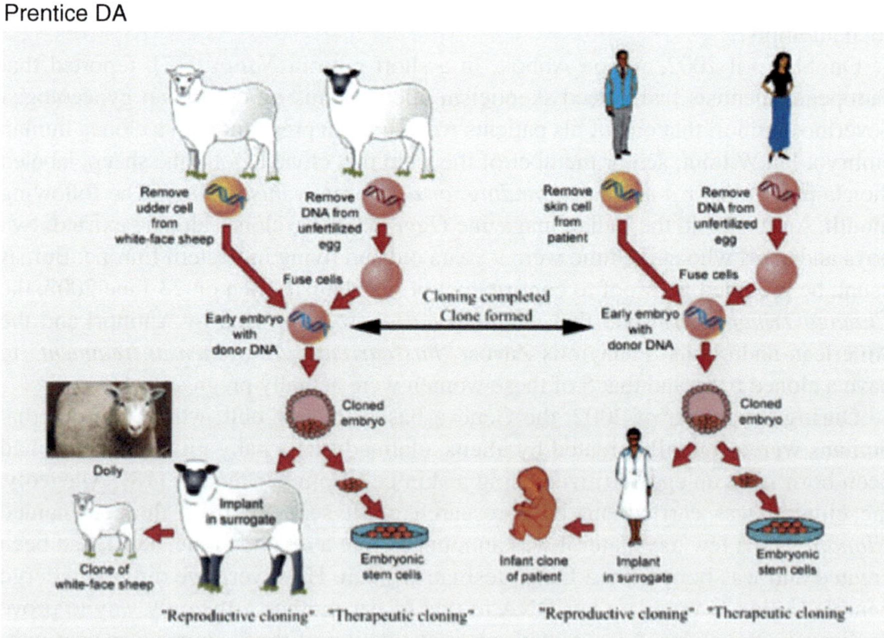

Fig. 8.2 Schematic representation of the procedure employed to clone the sheep Dolly (left portion of the figure). The same procedure would be employed when cloning a human being (right portion of the figure). (From Prentice DA, with permission)

documentation of the data of all of the accomplished results depicted herein will be described in detail in peer-reviewed journals" never materialized. The second article [21] seems to deal with the same procedure of the first, since it describes the preimplantation embryonic potential of fibroblasts from adult skin cells from an infertile man. The fibroblasts were fused with both enucleated bovine oocytes and human oocytes obtained from the wife. Oocytes reconstructed via somatic cell nuclear transfer were cultured *in vitro*. Of three reconstructed human oocytes, one developed to the four-cell stage and was subsequently transferred into the patient's uterus, but no pregnancy developed. Authors claimed this to be "*the first evidence of the creation and transfer of a human cloned embryo for reproductive purposes*."

In commenting Zavos results, Robert Edwards [22] cautioned: "*A wide perspective must be maintained on this work. Results in many animal species remain disastrous, as in mice, with many fetuses and offspring grossly malformed. Dolly has just died, seemingly prematurely. Results in cattle have greatly improved, with many embryos growing to full term. … Why do such immense differences arise between species*?". In a further comment, Azim Surani [23] stated that he was not convinced at all that the brief summary "*contains enough information to reach any valid conclusions. … The present paper is broad-brush treatment of a difficult subject that lacks attention to detail, and as such creates a false impression of the state of knowledge and efficiency of the procedure*." Surani in vain hoped that essential details would be forthcoming.

As mentioned, also in the case of HTC—i.e., the technique envisioned in order to generate matched nuclear transfer (NT)-ESCs by SCNT—initial reports did not stand up to scrutiny: In 2004 and 2005, two reports in the journal *Science* from a group led by the Korean researcher Hwang Woo-suk claimed to have obtained human ESCs by cloning [24, 25], but these experiments were shown to be fraudulent. In 2013 Tachibana et al. [26] identified a premature exit from meiosis in human oocytes coupled by a suboptimal activation as key factors responsible for failure of early attempts. By optimizing SCNT, they were able to circumvent these limitations, leading to successful derivation of human NT-ESCs displaying normal diploid karyotypes. These embryonic structures inherited their nuclear genome exclusively from parental somatic cells, and their gene expression and differentiation were similar to embryo-derived ESCs, suggesting an efficient reprogramming of somatic cells to a pluripotent state.

The following year, two teams generated embryonic stem cell lines by SCNT from adult human cells [27, 28], but there is at present a halt in work with this technique.

The Position of National and International Bodies Toward Human Reproductive Cloning

Over the years that followed the first claims of having achieved HRC, opposition to it grew, not only within the public at large but also from almost every existing institution, national and international. Requests for banning research in this field have been made in a number of countries, and as of 2021, some 45 countries have formally banned human cloning. Although no such prohibition exists at federal level in the USA, several individual states had done so.

At the international level, the World Health Organization (WHO) has been at the forefront of the campaign against HRC: as early as 1997, with a solemn and unanimous declaration, the World Health Assembly condemned any form of human cloning, affirming that "*the use of cloning for the replication of human beings is ethically unacceptable and contrary to human integrity and morality*" [29]. Intriguingly, the only member state opposing the declaration did so on the ground that it was not "strong enough." Opposition to HRC was reiterated by the WHO's Executive Board in 2005 [30].

Additional international documents followed the WHO's condemnation. Among them are as follows: the Universal Declaration on the Human Genome and Human Rights, adopted by the UNESCO (United Nations Educational, Scientific and Cultural Organization) General Conference in 1997 and endorsed by the United Nations General Assembly the following year, and the World Medical Association's Resolution on Cloning, approved in 1997.

Elaboration of an international convention against HRC has been under consideration in the United Nations since December 2001. Although all countries oppose the procedure, some favored a comprehensive ban to include also HTC; others wanted the ban to cover only HRC. Often, when members of intergovernmental organizations cannot agree on a form of binding international law, they can settle for a declaration, which is less demanding. This is what happened at the UN: After 4 years of debate, on March 2005 the UN General Assembly approved a Resolution calling on member states to "*adopt all measures necessary to prohibit all forms of human cloning, inasmuch as they are incompatible with human dignity and the protection of human life.*" The text was adopted by a vote of 84 in favor to 34 against, with 37 abstentions. The resolution also contained a call "*to protect adequately human life in the application of life sciences; to prohibit the application of genetic engineering techniques that may be contrary to human dignity; to prevent the exploitation of women in the application of life sciences; and to adopt and implement national legislation in that connection*" [31].

In 2008, UNESCO decided to add its voice to that of WHO and the UN, and its Bioethics Program began to investigate the possibility of a convention on cloning. There was tension between the independent experts supporting a ban on HRC, and member states concerned that disagreement would again surface, and ultimately the idea of a cloning convention dropped from UNESCO's agendas in 2012. The idea was taken up again in 2014, but despite a growing consensus, as of the end of 2021, there has been no move on the part of UNESCO to start to develop a treaty [32].

This situation is considered unsatisfactory because, for those states that have yet to formulate national regulations or policies on HRC, the absence of a clear international guidance may hinder an affirmative action.

The Inter Academy Partnership (IAP) is an umbrella organization comprising more than 140 national, regional, and global member academies, working together to support the role of science in seeking evidence-based solutions to global challenging problems. In 2013, IAP issued a statement calling for a ban on HRC while,

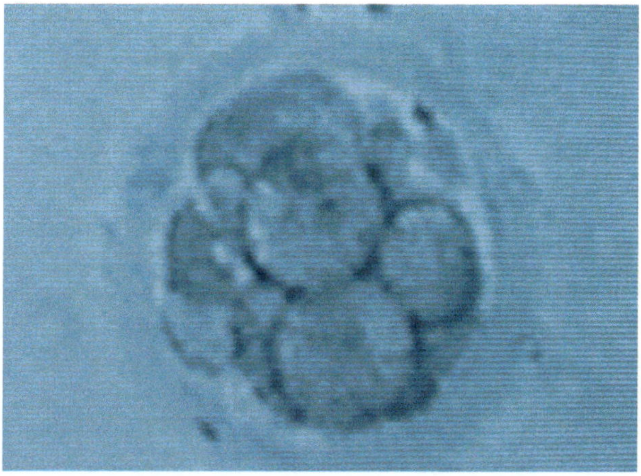

Fig. 8.3 An 8–10-cell human embryo derived from somatic cell nuclear transfer of granulosa, at 92 h. (From Zavos, 2003 [15])

at the same time, excluding from this ban cloning to obtain embryonic stem cells for both research and therapeutic purposes (see Fig. 8.3).

In conclusion, on the one hand, there has been widespread opposition to the cloning of a human individual on ethical grounds even before scientists started cautioning that there are major technical problems to be resolved; on the other, the way to an international treaty is still full of obstacles.

Ethical Considerations

An early publication by Savulescu [33] produced a list of arguments against or in favor of human cloning, bearing in mind that he mixed HRC and HTC.

On the negative side he listed

1. It is liable to abuse.
2. It violates a person's right to individuality, autonomy, selfhood, etc.
3. It violates a person's right to genetic individuality (whatever that is—identical twins cannot have such a right).
4. It allows eugenic selection.
5. It uses people as a means.
6. Clones are worse-off in terms of well-being, especially physiological well-being.
7. There are safety concerns, especially an increased risk of serious genetic malformations, cancer, or shortened lifespan.

Arguments in favor of HRC include

1. General liberty justifications.
2. Freedom to make personal reproductive choices.
3. Freedom of scientific inquiry.

4. Achieving a sense of immortality.
5. Eugenic selection (with or without gene therapy/enhancement).
6. Social utility—cloning socially important people.
7. Treatment of infertility (with or without gene therapy/enhancement).
8. Replacement of a loved dead relative (with or without gene therapy/enhancement).
9. "Insurance"—freeze a split embryo in case something happens to the first: as a source of tissue or as replacement for the first.
10. Source of human cells or tissues.
11. Research into stem cell differentiation to provide an understanding of aging and oncogenesis.
12. Cloning to prevent a genetic disease.

In commenting Savulescu's list, Williamson [34] stressed the importance of the right of a person to individuality, autonomy, and identity, associated with the right of not becoming a "means."

A substantial argument is that proposed by Jonas [35] who pointed out that technology requires developing an expanded conception of responsibility. Indeed, those with access to modern technologies raise the prospect of modifying our own genetic nature and significantly affect future generations. In this context, a cloned individual would be the victim of a clear violation of the basic right not to receive unrequested information about one's genetic status: A cloned young adult will inevitably know to have all genetic abnormalities he/she will see in the person with whom he/she shared the entire genome.

In order to evaluate the abovementioned arguments in a simple, yet unbiased manner, an essay published in 2002 [36] proposed to utilize two very simple criteria to evaluate reproductive cloning [37]:

1. Any technique aimed at producing an offspring who is the biological child of the *two* members of the couple involved must be—in principle at least—considered legitimate. This does not mean that every technique not fulfilling this criterion must be automatically labeled as unethical or—even worse, banned. Indeed, in a modern, democratic, and pluralistic society, where a number of ethical viewpoints coexist, legislation should only outlaw those techniques that are perceived as *causing damage*.

   It is easy to recognize that this is a concept difficult to define. Indeed, a technique could produce physical damage to the individual submitting to it, or to the offspring; it could also impact negatively on the psychological status of the individual born thanks to the technique, and it could have negative consequences for a community, or society at large.
2. Reproduction (irrespective of whether it is achieved through natural means or following medical assistance) must be considered a project aimed at giving birth to a new human being with rights that are identical to those of her/his parents, not a process to produce a child "at all costs."

> Such a principle was recognized very early in the ample debate that took place in the United Kingdom, the first country to be confronted with assisted reproduction. The *Code of Practice* of the *Human Fertilisation and Embryology Authority*, issued in its revised form in 1993, states clearly that one of its aims is "a concern for the welfare of the children, which cannot always be adequately protected by concerns for the interests of the adults involved" [38].

For those accepting these principles, as the international community seems to have done, HRC cannot be considered an important aid to people with no gametes who wish to reproduce, because—as mentioned—the technique contradicts the very basis of reproduction. First of all, reproductive cloning will not help couples wishing to have their own biological children, since in this case the offspring will only be the biological child of one parent, especially when a woman has no ovaries and therefore not even the maternally inherited cytoplasmic DNA can be passed to the child.

Finally, HRC allows the dominance of one human being (the nucleus donor) on the corporeal identity of another human being (the cloned one), representing a clear attempt at selecting the physical characteristics of a person, a fact contrary to the basic ethical principle of equality, since the clone will not have the opportunity to enjoy the diversity resulting from randomly inheriting its DNA from a man and a woman.

Based on these overarching principles, a number of bodies concerned with ethics have condemned HRC. In July 2002, the PCB, President's Council of Bioethics (established by the President of the USA), has forcefully addressed the issue of HRC affirming that proponents have defended their position "*by appeals to the good of freedom, existence* (*as opposed to non-existence*) *and well-being*." To this the PCB has responded that these arguments "*overemphasize the freedom, desires and control of parents and pay insufficient attention to the well-being of the cloned child-to-be*." They concluded: "*The Council holds that, once the child-to-be is carefully considered, these arguments are not sufficient to overcome the powerful case against engaging in cloning-to-produce-children*." The PCB unanimously reached the uniquely strong conclusion that rejection—on moral grounds—of cloning for human reproduction "*is not, as sometimes implied, a merely temporary objection, easily removed by the improvement of technique*." In fact, there are "*reasons for believing that the safety risks might be enduring*" and "*that conducting experiments in an effort to make cloning-to-produce children safer would itself be an unacceptable violation of norms of research ethics*." For this reason, "*there seems to be no ethical way to try to discover whether cloning-to-produce-children can become safe, now or in the future*" [39].

The ASRM issued a document in 2016 [1], stressing that "*reproductive SCNT* (somatic cell nuclear transfer) *has been inefficient in non-human species, with relatively few births reported in veterinary studies. It also has been associated with harmful complications in most mammalian species, including high fetal and neonatal death rates and/or imprinting and developmental disorders*." The document accepts the fact that technological progress can increasingly reduce complications, but—as stated by Robert Edwards [22]—"*Cloning is still a matter of argument about animals, where results in most, if not all, species so far cloned by nuclear transfer have been appalling. Perhaps no-one would accept moving to human studies while disasters, evidently due to imprinting, afflict virtually every cloned offspring.*" The ASRM document summarizes arguments against and in favor of HRC [1]: Opponents voice the fact that both natural conceptions, or one of the forms of assisted reproduction, involve the birth of an offspring with a uniquely mixed genetic lineage, whereas, following HRC, the offspring will have the genome of the donor of the somatic cell utilized. For this reason, no situation can justify recurring to it. On the opposite front: "*In the case of infertile couples in which one or neither partner can produce gametes, two situations might apply. If the male partner cannot reproduce with his spermatozoa, reproductive SCNT with his somatic cell would enable him to have a genetic tie with the child. His partner would have a biological tie if she donates the recipient oocyte or gestates the child. If the female partner cannot reproduce with her ova, transferring the nuclear DNA from her somatic cell to an enucleated donor oocyte would allow her to have a genetic relation to the child, although her partner would not.*"

In conclusion, the committee believes that "*As long as the safety of reproductive SCNT is uncertain and infertile individuals/couples have alternatives for conception, the application of reproductive SCNT by medical professionals does not meet standards of ethical acceptability.*" The document, however, leaves unanswered the main question posed by the PCB that it would be impossible to arrive in an ethical way at a safe way to use HRC.

Of interest is the statement that a negative outlook at HRC "*should not, however, be used to prohibit research in therapeutic SCNT, which can be ethically justifiable.*"

Conclusions

Two types of concerns have surfaced in the debate over human cloning: *safety* and *human rights*.

On the safety issue, it has been argued that widespread hostility is inherent in an illogical and, therefore, transient fear of every new technology. Indeed, it has been documented that human low fecundability is due to a fairly large rate of re-absorption of early, severely malformed embryo indicating the existence of mechanisms capable of recognizing and eliminating the vast majority of developmental errors [40]. In addition, proponents argue that with presently available diagnostic tools, even if errors occur, they can be easily identified and dealt with. This approach must be held unethical since it considers a new human being a "commodity," to be created and eliminated if defective.

On the ethical front, whereas it is true that, as scientific knowledge proceeds, ethical considerations may also evolve, the fact remains that—at present and for the foreseeable future—there seems to be no ethical way to investigate whether cloning-to-produce-children can become safe. More fundamentally, HRC deprives the new individual of one of its basic human rights, that of being born out of the diversity inherent in sexual reproduction.

References

1. American Society for Reproductive Medicine (ASRM) Ethics Committee Report. Human somatic cell nuclear transfer. Fertil Steril. 2016;105:e1–4.
2. Hong TK, Song J-H, Lee S-B, Do J-T. Germ cell derivation from pluripotent stem cells for understanding in vitro gametogenesis. Cell. 2021;10:1889.
3. Segers S, Guido Pennings G, Dondorp W, de Wert G, Mertes H. In vitro gametogenesis and reproductive cloning: can we allow one while banning the other? Bioethics. 2019;33:68–75.
4. Zafarris J. The etymology of "clone". https://uselessetymology.com/2018/05/16/the-etymology-of-clone/.
5. Campbell KH, McWhir J, Ritchie WA, Wilmut I. Sheep cloned by nuclear transfer from a cultured cell line. Nature. 1996;380(6569):64–6.
6. Rhind S, Cui W, King T, Ritchie W, Wylie D, Wilmut I. 69 Dolly: a final report. Reprod Fertil Dev. 2004;16:156.
7. Shiels PG, Kind AJ, Campbell KH, Waddington D, Wilmut I, Colman A, Schnieke AE. Analysis of telomere lengths in cloned sheep. Nature. 1999;399:316–7.
8. Sinclair KD. Dolly at 25… is she '… still goin' strong? Reproduction. 2021;162:E1–3.
9. Sinclair KD, Corr SA, Gutierrez CG, Fisher PA, Lee JH, Rathbone AJ, Choi I, Campbell KH, Gardner DS. Healthy ageing of cloned sheep. Nat Commun. 2016;7:12359.
10. Wilmut I, Bai Y, Taylor J. Somatic cell nuclear transfer: origins, the present position and future opportunities. Philos Trans R Soc Lond Ser B Biol Sci. 2015;370(1680):20140366.
11. Tsunoda Y, Shioda Y, Onodera M, Nakamura K, Uchida T. Differential sensitivity of mouse pronuclei and zygote cytoplasm to Hoechst staining and ultraviolet irradiation. J Reprod Fertil. 1988;82:173–8.
12. Simerly C, Dominko T, Navara C, Payne C, Capuano S, Gosman G, Chong KY, Takahashi D, Chace C, Compton D, Hewitson L, Schatten G. Molecular correlates of primate nuclear transfer failures. Science. 2003;300(5617):297.
13. Noggle S, Fung H-L, Gore A, Martinez H, Crumm Satriani K, Prosser R, Oum K, Paull D, Druckenmiller S, Freeby M, Greenberg E, Zhang K, Goland R, Sauer MV, Leibel RL, Egli D. Human oocytes reprogram somatic cells to a pluripotent state. Nature. 2011;478(7367):70–5.
14. Kim K, Kenigsberg S, Jurisicova A, Bentov Y. The role of mitochondria in oocyte and early embryo health. OBM Genetics. 2019;3:29.
15. Haldane JBS. Biological possibilities for the human species of the next ten thousand years. In: Wolstenholme G, editor. Man and his future. Boston: Little, Brown and Co.; 1963.
16. Writer S. In the news: Antinori and Zavos. Times Higher Education. www.timeshighereducation.co.uk/164313.article. Accessed 10 Aug 2001.
17. Abbot A. Disbelief greets claim for creation of first human clone. Nature. 2002;416:570.
18. Cult scientists claim first human cloning. The Guardian. https://www.theguardian.com/science/2002/dec/28/genetics.science. Accessed 28 Dec 2002.
19. Coghlan A. Latest human cloning claims leave sour taste. https://www.newscientist.com/article/dn17002-latest-human-cloning-claims-leave-sour-taste/. Accessed 22 Apr 2009.
20. Zavos PM. Human reproductive cloning: the time is near. Reprod Biomed Online. 2003;6:397–8.
21. Zavos PM, Illmensee K. Possible therapy of male infertility by reproductive cloning: one cloned human 4 cell embryo. Arch Androl. 2006;52:243–54.
22. Edwards RG. Human reproductive cloning a step nearer. Reprod BioMed Online. 2003;6:399–400.
23. Surani A. False impressions on human cloning. Reprod Biomed Online. 2003;6:398–9.

24. Hwang WS, Ryu YJ, Park JH, Park ES, Lee EG, Koo JM, Jeon HY, Lee BC, Kang SK, Kim SJ, Ahn C, Hwang JH, Park KY, Cibelli JB, Moon SY. Evidence of a pluripotent human embryonic stem cell line derived from a cloned blastocyst. Science. 2004;303(5664):1669–74.
25. Hwang WS, Roh SI, Lee BC, Kang SK, Kwon DK, Kim S, Kim SJ, Park SW, Kwon HS, Lee CK, Lee JB, Kim JM, Ahn C, Paek SH, Chang SS, Koo JJ, Yoon HS, Hwang JH, Hwang YY, Park YS, Oh SK, Kim HS, Park JH, Moon SY, Schatten G. Patient-specific embryonic stem cells derived from human SCNT blastocysts. Science. 2005;308(5729):1777–83.
26. Tachibana M, Amato P, Sparman M, Marti Gutierrez N, Tippner-Hedges R, Ma H, Kang E, Fulati A, Lee H-S, Sritanaudomchai H, Masterson K, Larson J, Eaton D, Sadler-Fredd K, Battaglia D, Lee D, Wu D, Jensen J, Patton P, Gokhale S, Stouffer RL, Wolf D, Mitalipov S. Human embryonic stem cells derived by somatic cell nuclear transfer. Cell. 2013;153:1228–38.
27. Yamada M, Johannesson B, Sagi I, Cole Burnett L, Kort DH, Prosser RW, Paull D, Nestor MW, Freeby M, Greenberg E, Goland RS, Leibel RL, Solomon SL, Benvenisty N, Sauer MV, Egli D. Human oocytes reprogram adult somatic nuclei of a type 1 diabetic to diploid pluripotent stem cells. Nature. 2014;510:533–6.
28. Chung YG, Eum JH, Lee JE, Shim SH, Sepilian V, Hong SW, Lee Y, Treff NR, Choi YH, Kimbrel EA, Dittman RE, Lanza R, Lee DR. Human somatic cell nuclear transfer using adult cells. Cell Stem Cell. 2014;14:777–80.
29. World Health Organization. Cloning in human reproduction. Fiftieth world health assembly. Geneva: World Health Organization; 1997.
30. World Health Organization, Executive Board. Reproductive cloning of human beings: status of the debate in the United Nations General Assembly: report by the Secretariat. https://apps.who.int/iris/handle/10665/20263.
31. United Nations. General assembly adopts United Nations declaration on human cloning by vote of 84-34-37. New York: United Nations; 2005.
32. Langlois A. The global governance of human cloning: the case of UNESCO. Palgrave Commun. 2017;3:17019.
33. Savulescu J. Should we clone human beings? Cloning as a source of tissue for transplantation. J Med Ethics. 1999;25:87–95.
34. Williamson R. Human reproductive cloning is unethical because it undermines autonomy: commentary on Savulescu. J Med Ethics. 1999;25:96–7.
35. Jonas H. Technology and responsibility: reflections on the new tasks of ethics. In: Sandler RL, editor. Ethics and emerging technologies. London: Palgrave Macmillan; 2014. p. 37–47.
36. Benagiano G, Primiero FM. Human reproductive cloning. Int J Obstet Gynecol. 2002;79:265–8.
37. Benagiano G. Reproductive strategies for human survival. Proceedings international conference "Robert G. Edwards at 75". Reprod Biomed Online. 2002;4(suppl. 1):72–6.
38. Human Fertilisation and Embryology Authority. The embryo and fertilisation authority of the United Kingdom: code of practice. London: The HFEA; 1993.
39. The President's Council on Bioethics. Human cloning and human dignity: an ethical enquiry. Washington, DC: President's Council on Bioethics; 2002.
40. Benagiano G, Farris M, Grudzinskas G. The fate of fertilised human oocytes. Reprod Biomed Online. 2010;21:732–41.

The Postmenopausal Mother

9

Johannes Bitzer

Introduction

The most famous woman becoming a mother at late post-reproductive age was Sarah, thus fulfilling the wish and order of God and Abraham:

"Sarai, Abram's wife, had borne him no children. But she had an Egyptian slave named Hagar;

So she said to Abram, 'The Lord has kept me from having children. Go, sleep with my slave; perhaps I can build a family through her.'

Abram agreed to what Sarai said.

So after Abram had been living in Canaan 10 years, Sarai his wife took her Egyptian slave Hagar and gave her to her husband to be his wife……

……the promise of God to Abraham was fulfilled.

'Is anything too hard for the LORD? At the appointed time I will return to you, about this time next year, and Sarah shall have a son' [1].

The solution was based on God's given nature of female fertility.

In postmodern times, reproductive medicine offers new solutions.

Women can become pregnant almost without age limits.

A typical news in the new book of books (Internet) reads like this [2]:

"….But nature has been overruled. It has recently been announced that a 63-year-old woman gave birth late last year to a healthy baby girl. A doctor implanted into her hormonally primed uterus an embryo created in a test tube with her husband's sperm and a young donor's egg.

The woman's doctor, Dr. Richard J. Paulson, said she had lied about her age to get around his age limit of 55 years for in vitro fertilization.

J. Bitzer (✉)
Department of Obstetrics and Gynecology, University Hospital Basel, Basel, Switzerland
e-mail: Johannes.Bitzer@usb.ch

J. G. Schenker et al. (eds.), *Hot Topics in Human Reproduction*, Reproductive Medicine for Clinicians 3, https://doi.org/10.1007/978-3-031-24903-7_9

The 63-year-old woman was not the first postmenopausal woman to have a baby, only the oldest. In the last several years, progressively older women have given birth through in vitro fertilization.

So we now must contemplate the curious possibility of women on Medicare becoming pregnant" [2].

There is already a list of records of the oldest woman becoming pregnant which is led for the moment by a lady who got pregnant through reproductive medicine technology at the age of 70 years in India.

These citations point already to the various scientific, cultural, philosophical, and ethical discourses around this subject depending on different perspectives

- Man takes the place of God or his angels.
- Man overrules nature and natural laws and limitations.
- Men and in this case women hide their age and misinform doctors, or doctors act in hidden places.
- Natural and social life course concepts (age and aging) are put aside.

We are confronted with

- The normative question regarding the transgression of borders and limits given to us by nature (are we allowed to do what is feasible) [3, 4].
- The empirical questions about risks and benefits of such a transformation (transgression) for the individuals involved and the society as a whole [3, 4].

The Normative Question

The basic normative question here is: In which situations **of individual or collective suffering** should doctors (healthcare professionals) **not act although** methods of diagnosis and therapy are available and could be used?

There is no "objective" or scientific answer to this question, but as a normative question, the answer will depend on the underlying concepts about the world and human existence.

With those who believe in a preexisting stable and enduring order given by God, by nature, or a global mind (idealism) to which humans are subdue and should follow, they will view medical interventions to overcome post-reproductive infertility as opponent to the given order and therefore potentially destructive and dangerous (Hybris, Sin, Blasphemy).

This position is the basis for warnings about the belief that for every problem humans may have, there is a technological or scientific solution which may lead to disastrous consequences [3, 4].

The alternative view sees humans as constantly progressing and transforming themselves and their environment. There are no eternal and stable laws, but it is up to the humans to define their aims and their values depending on the specific environment they live in and in which the individual should have the opportunity to decide (autonomy).

The basis for this decision is then a question of benefits and risks and thus accessible to objective research (empiricism). There are no absolute a priori values.

Based on this autonomy (self-definition), the responsibility regarding the consequences of the decision to become a mother in the post-reproductive phase rests with the individual (freedom of will) [3, 4].

In medicine this approach has been operationalized in the bioethical principles of Beauchamp and Childress [5]

- Respect for the autonomy
- Non-maleficence; do not harm
- Beneficence; do good and promote health
- Justice; make no differences in providing help

Bioethics of Late Motherhood

The postmenopausal woman wants to become pregnant and have a child and she asks for medical help.

The reason for this request is not a disease. She is suffering from an unfulfilled personal wish, an aim in life she cannot attain without medical help.

In other words, the woman is suffering from her loss of her fertility.

Is it the task of medicine to restore lost capacities necessary for the individual to feel good, to achieve a goal in life and give back a meaning to life [6, 7]?

The answer is a clear yes. In many fields of medicine, the interventions aim at restoring lost functions and improve quality of life (from orthopedics to psychotherapy, from sexual medicine to rehabilitation) [8].

According to WHO, health does not only mean the absence of disease but also physical and mental well-being and quality of life including the attainment of goals (especially regarding sexual and reproductive health) [9].

The special issue in this situation is however that the wish, the goal, and the restoration of fertility aims include another human being thus going beyond the person asking for help. This human being (the desired child) does not yet exist, cannot be included into the decision-making process, and thus presents a special challenge to which there are several possible answers [10].

In one perspective, this not (yet) existing being cannot be the subject of the ethical principles mentioned above. There is no person with autonomy, and therefore questions of harm or beneficence or justice are not applicable for the medical team.

The woman alone has the autonomy and is object of the principle of no harm. It is the mother's responsibility to ensure no harm and beneficence for the child to be.

In the alternative perspective, the desired child is integrated into the ethical consideration taking into account the complete dependency from the mother and thus focusing on the environment into which the child is born.

Applying bioethical principles to both—the mother and the child to be—can help clarify the role, tasks, and responsibilities of reproductive medicine in the context of the pregnancy and childbirth in postmenopausal women.

The Woman

Her wish for a child is an essential part of her reproductive autonomy. She decides when and how many children she wants to have with a partner of her choice.

Giving the individual the autonomous decision about reproduction is a central element of reproductive rights [9].

In the respective laws, the focus is on the protection against interventions from outside—like forced sterilization, prohibition of contraceptives, etc.

Reproductive medicine focuses more on overcoming biological barriers to reproductive aims.

The woman has independent of her age the right to get help for realizing her reproductive autonomy. It follows from the autonomy principle that the medical professionals will do their best to achieve a pregnancy [3–5].

What About the Second Principle? Do Not Harm

There are no studies indicating that late motherhood would increase maternal mortality significantly neither during the pregnancy nor during delivery. There is nonetheless an age-related risk increase for complications like gestational diabetes, preeclampsia, etc.

In one study among women in their sixth decade of life [11], 35% experienced pregnancy-induced hypertension, 20% developed gestational diabetes, and 78% underwent a cesarean section [11]. The risks were even higher in women more than 55 years old, compared with those 50–54 years old.

The higher complication rate can however be largely compensated, thus making postmenopausal age not a contraindication to the use of modern reproductive technology [12].

The same is true for pregnancy independent morbidity and mortality which does not seem to be significantly different between pre- and postmenopausal women at least until the age of 60 [12].

This means that we can assume that the life expectancy of pregnant women in the post-menopause is more or less the same as in the same age cohort. Life expectancy is thus not an issue for the woman but possibly for the child to be born (see below).

It is however important to stress the point that information about risks is based on scientific evidence and it is still the individual woman who should be enabled to give individual weight to the numbers (shared decision-making).

What About Mental Health?

Postmenopausal women have a higher risk to develop a postpartum depression [13].

Theoretically (there is a lack of empirical evidence), it could be assumed that there is an age-related decrease in the resilience to stress.

During the aging process, the capacity to fulfill different tasks at the same time declines.

At the same time, the frequency of mood changes and anxiety seem to increase [13].

Another aspect which has received not so much attention in research is the concept of age-related developmental steps or tasks and related life phase developmental crisis.

In general, we differentiate between different so-called life phases which follow one after the other: childhood, puberty and adolescence, adult life, and aging [14, 15].

The early phases of human development have been extensively explored, and it is generally agreed upon that the development from the child to the adult is an upward dynamic with an increase in competencies (cognitive behavioral, social, etc.) [15].

The aging process is considered to be a dynamic characterized either by general decrease in some of these capacities (performances) with a possible shift to other qualities (maturation perspective) with the final point of death.

In the developmental model, the human being fulfills specific steps or better tasks of development whereby the successful mastering of one step has an important influence on the following one [15].

The child and adolescent psychology and psychiatry has provided a huge amount of knowledge and insight which has become part of our general understanding of these life phases [15].

Regarding the so-called second phase of this developmental curve, there are different models of understanding the questions about the challenges and responses regarding aging and the respective "developmental tasks" [15].

A frequently cited model is the one described by Erikson [16].

Erikson differentiates the reproductive phase from the two post-reproductive phases whose developmental tasks are generativity and integrity.

The previous developmental goals regarding identity, work, and family are getting less important, and new tasks emerge like summarizing life experience and to give knowledge and the work done to others (creativity and generativity), the reflection about age and death, the limitations of time, and potency leading to an acceptance and inner serenity (integrity).

Taking this as a more general background, the question can arise whether these developmental tasks in the confrontation with aging, the limits, and limitations are eventually simply put aside and hinder thus the maturation process of the individual.

From a theoretical point of view, this "moving back in time" might have a different impact on different women with differences in their personality and life circumstances but especially also in their reproductive biography.

The woman who already has children is in a different position compared to the nulliparous woman.

In line with this, the post-reproductive phase can be considered the phase of grand-parenthood, the phase in which there is less responsibility for and involvement with the small completely dependent child thus creating room for a different relationship (the grandparent as the admiring and supporting companion liberated from the everyday stress).

The postmenopausal mother who already has adult children creates a new family structure with new social interactions and role definitions which demands adaptions and flexibility in the family system which may be perceived as continuous stressors.

To summarize, it can be stated that based on the present knowledge the medical assistance in fulfilling the wish for a child of a postmenopausal woman carries increased risks during pregnancy for her physical health, but that modern obstetrics can in general manage these risks according to studies up to the age of 55–60 years this preserving her physical health.

The possible risks for her psychosocial health are not well explored and researched until now. This will be an important task for the future; on one hand, it will help to counteract possible myths and prejudices (older-age is per se a psychological or psychiatric contraindication), and on the other hand, it can help the woman to evaluate her individual risks in relation to late motherhood.

The Child To Be

Including the "child to be" into the ethical discourse, the basic fact which has to be taken into account is the complete dependency of the child on the care by others, this radical external determination being completely at the mercy of the mother or others [17].

For all reproductive interventions, be it contraception, pre-conceptional genetic test, the preimplantation and prenatal diagnosis, the donation of germ cells, or surrogate motherhood, the common denominator is the fact that the involved parents may have obtained more freedom regarding their will and decisions but they do not change the radical lack of freedom of the child which continues during the early years of life.

The child cannot warm itself; it must be warmed; it cannot feed itself but must be fed; it cannot move by itself but must be moved [17].

The child is put into a network of human relationships outside of its own control or the capacity to create such a network by itself.

Therefore, we cannot assume any sort of autonomy of the "child to be."

This lack of interaction with an autonomous partner poses the question of the ethical-moral status of the child to be and which rights for the future child can be derived from that [17, 18].

In early stages of development, laws have mainly the function to protect the embryo and the newborn from external damage and interventions [19]. Society or better the state exerts a sort of "negative" duty of care by issuing rules and regulations aiming to ensure the physical integrity of the embryo and the newborn, thus assuming that the embryo and the fetus would have the autonomous wish to be healthy and to survive (an ontological autonomy).

This concept is the base also for the "in the future projected" autonomy, for example, for intersex babies who cannot decide about interventions on their body when they are born but should have the autonomy to decide at later age (as adolescents or adults), thus prohibiting early sex determinant interventions.

This protective approach is mainly focusing on the do not harm principle, but is not adequately addressing the principle of beneficence for the child to be.

The principle of beneficence poses a challenge to all those involved in bringing this child into life [20].

The more we intervene and modify the process of coming into the world of this human being completely depending on a network of relationships, support, and love, the more the question comes up whether the newborn child has some inborn right to be born in an environment that provides safety and allows growing and flourishing, in other words "good and responsible parenting" [20].

But what is "good parenting?" Are those external norms and concepts not discriminating against the autonomy of the couple who needs help to achieve a pregnancy and have a child and who has the right to evaluate these couples with respect to their "parenting?"

This tension and possible conflict between the respect for the autonomy of the future parents and the beneficence duty toward the future child is up till now not resolved with generally agreed guidelines.

Three basic requirements for ensuring the future well-being and psychosocial health of the child seem to be agreed upon [21–24].

Health Conditions and Life Expectancy of the Parents

This relates to the presence and availability of "mother" and "father" roles (physical and psychological). Those are the persons having responsibility for the child. There is some uncertainty how this age limit should look like. But there is some agreement that a probable time span of 20 years regarding the availability of the parents is desirable.

Stressful Life Conditions and Resilience of the Parents

Small children need affection, time, and patience, being prepared to put one's own interests behind and the needs of the child in front.

"Parenting, the most complicated job in the world" (Virginia Satir) [25].

The interpersonal skills are empathy, positive encouragement, and feedback, serving as a role model, helping the child to regulate emotions (esp. negative emotions), helping the child to cope with failures and getting disappointed, becoming aware of self-harming behavior, establishing relationships and learning social behavior, development of cognitive skills, etc.

Future parents who live under conditions of chronic stress (physical, psychological, social) or those who suffer from affective disorders or other psychological problems leading to a reduction of their capacity to care are in danger to not being able to provide the necessary conditions for a healthy development of the child.

Resources and Family Structures of the Future Parents

Besides the stressors, resources play an important role for the environment in which the child is born. Resources include other family members and other individuals in the environment of the child who can provide reliable availability and allow the establishment of relationships, interpersonal learning, and encouragement.

These three dimensions may serve as an orientation regarding the ethical question of the well-being (Kindeswohl) of the child to be.

The realization of the ethical duties of not doing harm to the child and of promoting health will depend on the balance of these dimensions.

References

1. Mose 18 Vers 14.
2. www.nytimes.com/1997/04/25/.../pregnant-at-63-why-not.html.
3. Beauchamp TL. Philosophical ethics. An introduction to moral philosophy. New York: McGraw Hill; 2001.
4. Hinman L. A pluralistic approach to moral theory. 5th ed. Boston: Cengage Learning; 2012.
5. Beauchamp TL, Childress JF. Principles of biomedical ethics. Oxford: Oxford University Press; 2001.
6. Cutas D. Postmenopausal motherhood: immoral, illegal? A case study. Bioethics. 2007;21(8):458–63.
7. Shufaro Y, Schenker JG. Pregnancies beyond the human biological fecundity. Womens Health (Lond). 2012;8(1):49–55.
8. Ekberg ME. Assisted reproduction for postmenopausal women. Hum Fertil (Camb). 2014;17(3):223–30.
9. WHO. www.who.int/about/governance/constitution.
10. Larcher V. Ethical issues in respect of children born after assisted reproduction technologies. Arch Dis Child. 2007;92:670–1.
11. Paulson RJ, Boostanfar R, Saadat P, Mor E, Tourgeman DE, Slater CC, et al. Pregnancy in the sixth decade of life: obstetric outcomes in women of advanced reproductive age. JAMA. 2002;288:2320–3.
12. Harrison BJ, Hilton TN, et al. Advanced maternal age: ethical and medical considerations for assisted reproductive technology. Int J Womens Health. 2017;9:561–70.
13. Naomi J. Spence: The long-term consequences of childbearing: physical and psychological well-being of mothers in later life. Res Aging. 2008;30(6):722–51.
14. Bitzer J, Horne AW. A new aged has come: the redefinition of women's health care. J Fam Plan Reprod Health Care. 2012;38:68–9.

15. Alder B, Abraham C. Psychology and sociology applied to medicine. 3rd ed. Amsterdam: Elsevier; 2009.
16. Erikson EH. The life cycle completed. New York: W. W. Norton & Company; 1998.
17. Wiesemann C. Fortpflanzungsmedizin und die Ethik der Elternschaft. J Reproduktionsmed Endokrinologie. 2007;4(4):189–19.
18. Krones T, Richter G. Kontextsensitive Ethik am Rubikon. In: Düwell M, Steigleder K, editors. Bioethik. Eine Einführung. Frankfurt: Suhrkamp Verlag; 2003. p. 238–45.
19. Keller R, Günther HL, Kaiser P. Embryonenschutzgesetz. Kommentar zum Embryonenschutzgesetz. Stuttgart: Kohlhammer Verlag; 1992.
20. Wiesemann C, Von der Verantwortung C. ein Kind zu bekommen. Eine Ethik der Elternschaft. Munich: C. H. Beck Verlag; 2006.
21. Wiesemann C. Natalität und die Ethik von Elternschaft und Familie. Zeitschrift für Praktische Philosophie. 2015;2(2):213–36. www.praktische-philosophie.org
22. Braverman AM. Old, older and too old: age limits for medically assisted fatherhood? Fertil Steril. 2017;107(2):329–33.
23. Steiner AZ, Paulson RJ. Motherhood after age fifty: an evaluation of parenting stress and physical functioning. Fertil Steril. 2007;87:1327–32.
24. Daar J, Benward J, Collins L. Oocyte or embryo donation to women of advanced reproductive age: an Ethics Committee opinion. Fertil Steril. 2016;106:e3–7.
25. www.ronitbaras.com/family-matters/parenting-family/best-family-quotes-by-virginia-satir/.

Sex Preselection

10

Gamal Serour

Sex selection is the term currently used when the woman with or without her partner tries to dictate the sex of the baby to be born. All children regardless of the beliefs or desires of their parents should enjoy equal legal and social status and full support and love from their parents without discrimination on the basis of sex.

The parenteral desire to choose the child's sex dates from antiquity. The Jewish Talmud advised couples on means to favor the births of male or female children [1].

The Arabs in the pre-Islamic era more than 1400 years ago used to practice infanticide for sex selection. The Holy Qur'an condemned this practice. It says "On the Judgment Day the entombed alive female infant is asked, for what guilt was she made to suffer infanticide" [2]. The Holy Qur'an described the behavior of some fathers when they are told of the birth of their female child "….His face gets dark and he chokes with suppressed agony. Hiding himself from his people being ashamed of the bad news and wonders whether to keep her in hardness or bury her in the dust, how evil is their decision" [3].

Methods for sex selection of the child are rooted in folklore including positioning during intercourse, timing of intercourse in the menstrual cycle, vaginal douching, or intake of certain foods to enhance the conception in one sex or another. In Europe, interest in sex selection is not new. In Europe, for many countries the ability to choose the sex of one's children has been desired by couples [4]. Aristotle advised sexual intercourse in northern wind to get a boy and in southern wind to get a girl. In Germany, a father was advised to take an ax to bed with him if he wanted to conceive a boy [5]. In France, women desiring a boy were provided a diet rich in potassium and sodium and poor in magnesium and calcium concentration [6].

G. Serour (✉)
International Islamic Center for Population Studies and Research, Al Azhar University, Cairo, Egypt
e-mail: giserour1@link.net

J. G. Schenker et al. (eds.), *Hot Topics in Human Reproduction*, Reproductive Medicine for Clinicians 3, https://doi.org/10.1007/978-3-031-24903-7_10

Millot, the obstetrician of Queen Marie Antoinette of France [7], indicated that it is the last movement of the woman that determines the sex of the child. It is the side on which she lies at ejaculation time that drives the sex of the child: always a boy when she is on the right side and always a girl when she lies on the left side [7]. More scientific means includes the observation that conception close to ovulation, which can be timed by measuring hormones, ultrasonography, or cervical mucus, is more likely to produce boys [8]. This chapter shall discuss sex selection in different cultures, the current available methods of sex selection, and its indications, guidelines, and ethical and legal concerns surrounding sex selection.

Sex Selection Practices in Different Cultures

The desire for sex selection is a reflection of cultures, tradition, religion, civilization, education, and available medical technologies in a given society or community, all of which may influence the morality and mentality of the people in these societies and communities. Interest in sex selection has a long history dating to ancient cultures [9]. Ancient Egyptians believed that women of certain complexion were destined to have boys. Early Greeks believed that tying off the left testicle would produce boys as the male determine sperm were derived from the right testicle.

The Babylonian Talmud advises couples on means that favor the birth of either male or female child (Niddah). The Hebrew Talmud suggested that placing the marriage bed in a north–south direction favored the conception of boys [10].

Arabs more than 1400 years ago, before Islam, used to practice infanticide for sex selection. The Holy Qur'an described this act and condemned it [2, 3].

In China female infanticide was mentioned in the historical records [11]. Sons are important for religious reasons. Chinese culture considers the family as a filial obligation. Only the names of the sons and grandsons are put in the list of the family genogram in the ancestral halls. The Chinese culture considers that the greatest sin of all is the death of a man without having a male son [12]. In old China the traditional cultural norm was to unilaterally divorce women who do not bear a son [13].

In India having more than one daughter was considered as a curse. Sons were important as they provided support for their aging parents [14]. Until recently, daughters did not have a right to inherit any part of the ancestral property of the parents. The prospective husband or his parents among the vast majority of Hindus generally demanded substantial dowries, which created lots of financial and psychological pressures on the wife to be or her family to fulfill their obligations. Recently the country prohibited this practice. Thus in the Indian culture, sex selection was bluntly misused and resulted in huge discrimination against the female child. The proportion of female to males had dropped from 935/1000 in 1981 to 927/1000 in 1991 and in some communities in the northern states of Bihar and Rajasthan to 600/1000 one of the lowest in the world [15].

In Japan the head of the family was always a man who controlled all family affairs. The first son held the right to succeed his father under legal protection. The low fertility rate in Japan tempted couples to select the sex of the child. Paradoxically the increasing average longevities of men and women in Japan encouraged parents' preference for female offspring to be better taken care of later in life [16].

Methods of Sex Selection

The traditional methods are far from being accurate and have a poor accuracy rate. With the rapid development of technology, more modern techniques became available with a high success rate in choosing the sex of the child. It is now possible to attempt to choose the sex of the baby before fertilization by separating the X and Y chromosome bearing sperm based on the 2.8% difference in their DNA content, using Microsort sperm separation technique. Embryos can be created by intrauterine insemination, IVF, or ICSI. Discarding disfavored sperm is less contentious than sex selection-based abortion and embryo wastage [17–20]. More recently the sex of the embryo could be selected in IVF programs by preimplantation genetic diagnosis (PGD) during IVF or ICSI [21–26].

In (PGD) the sex of the embryo can be diagnosed by using fluorescent in situ hybridization (FISH) or more recently comprehensive chromosomal survey (CCS). FISH identifies only a limited number of chromosomes including X and Y chromosomes by examining one or two cells from a cleaved embryo. CCS ensures transferring euploid embryos only and discard abnormal embryos. CCS examines all the 23 pairs of human chromosomes by examining 5–10 cells from day 5 or day 6 embryo blastocysts. These techniques add more cost to the already high expenses of IVF/ICSI.

CCS adds to sex selection the ability to analyze, select, and transfer to the mother only embryos of the desired sex that have the appropriate number and structure of chromosomes. Furthermore, single nucleotide polymorphism (SNPs) microarray which is associated with a host of physical traits had been proposed to optimize pregnancy outcome of the selected embryos by excluding embryos with gene defects or choosing embryos with specific characteristics.

Preimplantation genetic testing (PGT) for aneuploidy and sex selection using CCS have been proposed to improve IVF success rate. A prospective, randomized, multicentric, multinational study indicated that PGT for aneuploidy by CCS does not substantially increase the live birth rate in women aged 36–40 years and is associated with less cryopreserved embryos [27]. However, some other studies showed the opposite, and the jury is still there. PGT allows not only the choice of the sex of healthy embryos that do not suffer from genetic diseases but also may allow gene editing whether therapeutic to alleviate genetic diseases and pathological conditions or to enhance certain characteristics.

The slippery slope is when gene editing is used to enhance certain characteristics in the chosen embryos such as to achieve athletic success, more intelligence, artistical sensitivity, or talents [28].

Some ART clinics in Europe and the USA promote polygenic risk scores (PRS) as an add-on test on the selected embryos before transfer [29]. Genetic experts have stated that PRS testing of embryos is unusable, unethical, and impractical. PRS is unproven and unethical and prospected parents should be warned against such practices and given adequate and unbiased information [30].

Sex selection may be performed after occurrence of pregnancy by prenatal sex selection after identification of the sex of the baby by ultrasonography, amniocentesis, or chorionic villous sampling (CVS). Prenatal sex selection will necessitate termination of pregnancy if the undesired sex is diagnosed. This will raise an additional ethical, religious, and in some countries legal concerns.

Indications for Sex Selection

The indications of sex selection are broadly divided into medical and social. There are more than 350 sex-linked diseases in human [17]. Sex selection for medical indications prevents conception in an affected child, whether a male or a female child, and eliminates the birth of a diseased child. Some of the common sex-linked diseases are due to chromosomal abnormalities such as Turner syndrome, Klinefelter syndrome, and fragile X syndrome. Sex-linked diseases may be monogenic diseases due to a specific gene defect as a cause of the disease as cystic fibrosis, sickle all disease, hemophilia, and muscular dystrophy. The social indication of sex selection is to satisfy the desire of the prospective parents. It has been associated with a huge ethical debate, disapproval, and even condemnation and criminalization which varied from one country to another. The disapproval and condemnation are based mostly upon prejudice against the female child and to a lesser extent on the argument that it may lead to disturbance of the global female/male ratio as happened in the past in some regions in India and China.

Treating Different Cases as if They Are Alike in Sex Selection

Dickens et al. reported that in ethical and legal analysis, the principle of justice that like cases be treated alike receives considerable attention. Less attention is given to the ethical injustice of treating different cases as if they are alike and applying an approach to a problem that is appropriate in one setting to a different setting in which that problem does not exist [31].

In China and India, sex selection practices disclose significant son preference resulting in birth ratio imbalance between the two sexes. In India the national sex ratio was 933 females to 1000 males and only 927 females in the age group under 6 years in the year 2001 [32].

In China the issue of sex selection was complicated by the introduction of the one child policy. Couples in the urban areas were usually allowed to have one child. In the rural areas, couples whose first child was a girl were allowed to have a second child after a specific period of time [33]. This encouraged women who got pregnant in a female child to get rid of the female fetus and try again for a male child resulting in a serious imbalance in sex ratio in China [34]. Such policy has been revised recently because of decline in the fertility rate in China and the loss of the demographic dividend, which previously contributed to the enormous development in China during the past few decades.

In contrast many other countries do not have sex preference. A comprehensive survey in Canada found a large majority of Canadians do not prefer children of one sex or the other. The survey showed that virtually all prospective parents want and feel strongly about having at least one child of each sex [35]. With the recent decline in fertility rate, delay of age of mothers at first child birth [36], and increased expenses of living, such preference of one child of each sex may not be the same today.

In the USA, 90% of couples with two or three children and wanting only one more employed sex selection for the purpose of family balancing [17]. Interesting in both the USA and UK, over half of surveyed couples' selecting their children's sex choose girls [37, 38]. In a survey conducted in Germany, 58% of respondents stated no interest in their children's sex, 30%wished to have an equal number of girls and boys, and 92% found this practice to be unthinkable [39].

In the Middle East, where the population is largely Muslims with Christian minority, women's dignity and her status in the society are often related to her ability to have children in general and particularly sons. Sex selection for social reasons is practiced with some guidelines to avoid discrimination against either sex [40].

Thus there are contrasts in the ethical approaches to sex selection in different countries. In countries where discrimination against the girl child is pervasive, dominant selection of the male child is likely to be practiced. In countries where there is less or no discrimination, couples do not prefer children of either sex. In the latter case, sex selection can be allowed to assist families that want children of both sexes and to fulfill reproductive autonomy of the couples after proper counseling. This raises the question of whether it is just to apply the same ethical and legal approaches to sex selection in these different circumstances [31].

Ethical and Cultural Issues in Sex Selection

The strongest objection to sex selection is its discrimination against birth of the girl child. Selection of a male child appears as a symbol and a cause of the inferior status of and discrimination against the girl child and perpetuates the devaluation of women.

The practice of sex selection, to detect severe sex-linked genetic disorders, using the modern technologies which does not involve abortion or discarding healthy embryos is widely accepted in modern laws and ethical assessments. More contentious but arguably tolerable is sex selection by a couple with one child or two or more children of the same sex, boys or girls, who wish to have only one more child of the other sex [41]. In such cases, sex selection is not based on societal or cultural discrimination against either sex. It satisfies the autonomy of couple's reproductive choice [42]. The procreative autonomy is the right of the person to freely choose his/her/their reproductive performance including his/her/their reproductive potentials.

However, reproductive autonomy is basically a personal decision, yet some would argue that it is not merely so [42]. Reproduction itself is a process which does not involve solely the person or the couple who makes the choice; it involves the other partner; the baby to born and its right for protection, respect, and non-discrimination; the family; the society; and the world at large [42, 43]. Dworkin defined a right of procreative autonomy as "a right of people to control their own role in procreation unless the state has a compelling reason for denying them that control [44]". The decision not to transfer in vitro created embryos is within the unfettered decision of the woman or the couple who produced the germ cells which created these embryos. While the woman is entitled to refuse to implant any embryo, the decision to select between embryos is constrained by mortality. The woman or the couple should not choose between embryos in ways that might constitute unfair discrimination against one sex or another. Such unfair discrimination is likely to occur if selection is performed for the first child in the family or performed in families who have children of both sexes or the choice of one sex only all the time. It is logically argued if contraceptive technology is widely practiced to prevent the conception of both boys and girls, why couples should be denied the use of their procreative autonomy to choose the sex of the baby to be born to increase the gender variety in their families and not for gender discrimination.

The universal prohibition of sex selection, which does not involve abortion, would itself risk prejudice to women in many present societies specially when birth of sons or daughters remain central to women's well-being. Family balancing can be acceptable, for instance, where a wife had borne three or four daughters or sons and it was in her and her family's best interest that another pregnancy should be her last for health, economic, or personal reasons. Employing sex selection in such cases to ensure the birth of a daughter or a son might then be approved, to satisfy a sense of religious or family obligation and to save the woman the increasingly health risk-laden pregnancies. If sex selection is not performed for such couple, it involves prejudice against the woman as she will try repeated pregnancies to fulfill her wish to have a child of the other sex whether a boy or a girl. In some societies, the risk may also be societal as not performing sex selection may lead to divorce, extramarital relation, or separation as the couple's desire is not being fulfilled within their marital relationship.

Almost 20 years ago, the author called for an urgent need for concerned international organizations to issue binding guidelines on sex selection as soon sex selection may become available on the counter, and we become faced with the outcome of its bizarre use. Regrettably this is happening today where many ART clinics in many countries perform sex selection for social indications without ethical guidelines which perpetuates the female child. Practices of sex selection intended to promote gender discrimination are unacceptable independent of cultural, religious, political, and societal demands [9].

Sex selection for medical reasons is universally approved as it alleviates the human suffering and improves the quality of life of the child to be born. Sex selection for social reasons is performed to increase the gender variety in the family and/or restore the sex ratio in the family for various reasons [31, 45–49].

Sex selection for social reasons is surrounded with a huge ethical, legal, religious, cultural concerns, reproductive autonomy, and human right issues. Discussion of sex selection for social indications should take into consideration all these issues before being judgmental on whether to approve or disapprove sex selection for social indications.

Healthcare providers and patients alike, in an era of globalization, move freely around the world. Thus it is common for healthcare providers to provide medical care in reproduction to couples with a different cultural, religious, and ethical background. Moral dilemmas which exist in certain practices as sex selection should not be assumed to be applicable to all [50, 51]. Global bioethics must respect the whole diversity of world views of ethics, both religious and non-religious [52].

Providing quality healthcare service in reproduction which is culturally sensitive and ethically sound to all requires healthcare providers and institutions to be aware of these different perspectives.

References

1. Epstein I, editor. Niddah 3:31A1: The Babylonian Talmud (Slokti Iw translator). London: Sonciono Press; 1948. p. 2017.
2. Holy Qur'an Sura Al Taqueer. 81:8–9.
3. Holy Qur'an Sura Al Nahl. 16:58–59.
4. Cohen J. Gender selection. Is there a European view? J Assist Reprod Genet. 2002;19(9):417–9.
5. Largey G. Reproductive technologies: sex selection. In: Reich WT, editor. Encyclopedia of bioethics. New York: The Free Press; 1978. p. 1440.
6. Stoolkowski J, Emmerich E. Influence de la nutrition minerale de la vached sur la repartition des sexes dans la descendance. Endocrinology. 1971;1971(32):3.
7. Millot JA. Cart de proceer les sexes a volunte. 4th ed. Edinburgh: Royal College of Physicians of Edinburgh; 1820.
8. Carson SA. Sex selection: the ultimate in family planning. Fertil Steril. 1988;50:16–9.
9. Serour GI. Transcultural issues in gender selection. In: Daya S, Harrison R, Kempers R, editors. Recent 18th world congress on fertility and sterility, Montreal 2004. Amsterdam: Elsevier; 2004. p. 2004.
10. Genesis. 46:15.
11. Li D. Preference for sons: past and present. China Popul Today. 1997;14(5):15–6.
12. Vermier YC. Marriage laws and customs of China. Hong Kong: Chinese University Press; 1996. p. 1966.
13. Liu NCA. Family law for the Hong Kong SAR. Hong Kong: Hong Kong University Press; 1999.
14. Wertz DC, Fletcher JC. Fatal knowledge. Prenatal diagnosis and sex selection. Hastings Center Rep. 1986;19(3):21–7.
15. Mudur G. Indian medical authorities act on antenatal sex selection. Br Med J. 1999;319:401.
16. Mori T, Watanabe H. Ethical considerations on indications for gender selection in Japan. J Assist Reprod Genet. 2002;19(9):420–5.
17. Fugger EF, Black SH, Keyvanfark K, Schulman JD, et al. Births of normal daughters after Microsort sperm separation and intra uterine insemination, in-vitro fertilization, or intracytoplasmic sperm injection. Hum Reprod. 1988;13(9):2367–70.
18. Johanson LA, Welch GR, Keyvanfar K, et al. Gender preselection in humans? Flow cytometric separation of X and Y spermatozoa for the prevention of X-linked diseases. Hum Reprod. 1993;8:1733–9.

19. Levinson G, Keyvanfar K, Wu JC, et al. DNA based X-enriched sperm separation as an adjunct to preimplantation genetic testing for the prevention of X-linked disease. Mol Hum Reprod. 1995;10:979–82.
20. Vidal F, Fugger EF, Blanco J, et al. Efficiency of Microsort flow cytometry for producing sperm populations enriched in X- or Y-chromosome haplotypes; a blind trial assessed by double and triple colour fluorescent in-situ hybridization. Hum Reprod. 1998;13:308–12.
21. Edwards RG, Gardner P, et al. Sexing of live rabbit blastocysts. Nature. 1967;214:576–7.
22. Griffin DK, Handyside AH, Harper JC, et al. Clinical experience with preimplantation diagnosis of sex by dual fluorescent in situ hybridization. J Assist Reprod Genet. 1994;11:132–43.
23. Harber JC, Coonen E, Handyside AH, et al. Mosaicism of autosomes and sex chromosomes in morphologically normal, monospermic preimplantation human embryos. Prenat Diagn. 1995;15:41–9.
24. Lissens W, Sermon K. Preimplantation genetic diagnosis: current status and new development. Hum Reprod. 1997;12(8):1756–61.
25. Saiki RK, Scharf S, Faloona F, et al. Enzymatic amplification of B-globin genomic sequences and restriction site analysis for diagnosis of sickle cell anemia. Science. 1985;230:1350–4.
26. The ESHRE Ethics Task Force, Shenfield F, Penning G, Devroey P, Sureau C, Tarlatzis B, Cohen J. Task force 5: preimplantation genetic diagnosis. Hum Reprod. 2003;18(3):649–51.
27. Verpoest W, Staessen C, Bossuyt PM, et al. PGT for aneuploidy by microarray analysis of polar bodies in advanced maternal age: a randomized clinical trial. Hum Reprod. 2018;33(9):1767–76.
28. Serour GI, Serour AG. The Islamic perspective: application of advanced reproductive technologies to screen human embryos during IVF. In: Sills ES, Palermo GD, editors. The human embryos and preimplantation genetic technologies. Ethical, social and public policy aspects. San Diego: Academic Press; 2019.
29. Kent A. Polygenic risk score for embryosis it ethical in Jass. Amsterdam: Elsevier; 2022. p. 17–27.
30. Farzono F, et al. The use of polygenic risk scores in pre-implantation genetic testing: an unproven, unethical practice. Eur J Hum Genet. 2021;30(5):493–5. https://doi.org/10.1038/s41431-021-01000-x.
31. Dickens BM, Serour GI, Cook RJ, Qiu R. Sex selection: treating different cases differently ethical and legal issues in reproductive health. Int Gynecol Obstet. 2005;90:171–7.
32. Office of the Registrar General India. Registrar General of India, provisional population totals, census of India 2001. New Delhi: Office of the Registrar General India; 2001. http://www.censusindia.net/resultsmain
33. Chan CLW, Yip PSF, Ng EHY, Ho PC, Chan CHY, Au JSK. Gender selection in China, its meanings and implications. J Assist Reprod Genet. 2002;19(9):426–30.
34. Shanghai: Evening News, reported in Indian Express, August 12, 1988.
35. Royal Commission on New Reproductive Technologies. Proceed with care: final report of the Royal Commission. Ottawa: Minister of Government Services Canada; 1993. p. 889–90.
36. UNFPA. The state of world population 208. The power of choice: reproductive rights and the demographic transition. New York: United Nations Population Fund; 2018.
37. Batzofin JH. XY sperm separation for sex selection. Urol Clin N Am. 1987;14:609–18.
38. Lui P, Rose GA. Social aspects of over 800 couples coming forward for gender selection of their children. Hum Reprod. 1995;10:968–71.
39. Dahi E, Beutel M, Brosig B, Hinsch KD. Preconception sex selection for non-medical reasons: a representative survey from Germany. Hum Reprod. 2003;18:2231–4.
40. Serour GI, Serour AG. The impact of religion and culture on medically assisted reproduction in the ME and Europe. Reprod Biol Med Online. 2021;43(3):421–33.
41. Dickens BM. Can sex selection be ethically tolerated? J Med Ethics. 2002;2002(28):335–6.
42. Serour GI. Reproductive choice: a Muslim perspective. In: Harris J, Holm S, editors. The future of human reproduction. Ethics, choice and regulation, vol. 200. Oxford: Clarendon Press; 2000. p. 191–202.

43. Serour GI. Family and sex selection. In: Healy DL, et al., editors. Reproductive medicine in the twenty-first century. Proceedings of the 17th world congress on fertility and sterility, Melbourne, Australia, vol. 2002. New York: The Parthenon Publishing Group; 2002a. p. 97–106.
44. Dworkin R. Life's dominion. London: Harper Collins; 1993. p. 148.
45. Cooke RJ, Dickens BM, Fathalla MF, editors. Sex selection—abortion in reproductive health and human rights, vol. 12. Oxford: Clarendon Press; 2003. p. 363–71.
46. Kilani Z, Haj HL. Sex selection and preimplantation genetic diagnosis at the Farah hospital. Reprod Bio Med Online. 2001;4(1):68–70.
47. Malpani A, Malpani A, Modi D. Preimplantation sex selection for family balancing in India. Hum Reprod. 2002;17:11–2.
48. Schenker JG. Gender selection: cultural and religious perspectives. J Assist Reprod Genet. 2002;19(9):400–10.
49. Serour GI. Attitudes and cultural perspective on infertility and its alleviation in the Middle East Area. In: Current practices and controversies in assisted reproduction. Report of a WHO meeting. Geneva: World Health Organization; 2002b. p. 41–9.
50. Alali K, Serour GI, Bagheri A. Challenges in Islamic bioethics. In: Bagheri A, et al., editors. Islamic bioethics current issues and challenges inter cultural dialogue in bioethics. Singapore: World Scientific; 2017.
51. Serour GI. What is it to practice good medical ethics? A Muslim's perspective. J Med Ethics. 2015;41:121–4. https://doi.org/10.1136/medethics-2014.102301.
52. Campbell AV. Global bioethics: dream or nightmare? Bioethics. 1999;13(314):183–90.

Planned Oocyte Cryopreservation: Social Aspects

11

Avi Tsafrir and Jordana Hadassah Hyman

Introduction

Globally, maternal age at first birth has risen significantly. Given that female age is inversely related to fertility, the delay in attempting to conceive results in an increase in the number of women seeking assisted reproductive technology at advanced reproductive age [1]. However, Assisted Reproductive Technology (ART) efficiency declines remarkably in this age cohort, mainly due to reduced oocyte quality and quantity. Therefore, most of these treatments are unsuccessful, causing significant distress and disappointment.

Fertility preservation has become a popular alternative to delaying one's attempt to conceive. While embryo cryopreservation has long been an integral part of ART, oocyte cryopreservation has lagged behind. Being the largest kind of cell in the human body, and the one with the largest content of water, the oocyte is particularly vulnerable to the formation of ice crystals during the freezing process. Ice crystals may damage intracellular organelles, including the meiotic spindle, and alter membrane permeability. These processes may lead to cell injury and death and reduce the survival of cryopreserved oocytes when thawed [2].

The poor survival rates of oocytes that are thawed after using the "slow freezing" technique, which was originally the standard practice in IVF labs, initially precluded extensive use of this clinical option [2]. Later, vitrification was introduced as a superior method for oocyte cryopreservation. This method involves rapid transition from liquid to glass-like stage and avoids formation of ice crystals which are detrimental to the oocytes. The first birth from a vitrified-thawed oocyte was first reported in 1999 [3]. Further advances followed in the next decade and established

A. Tsafrir (✉) · J. H. Hyman
IVF Unit, Department of Obstetrics and Gynecology, Faculty of Medicine, Hebrew University of Jerusalem, Shaare Zedek Medical Center, Jerusalem, Israel
e-mail: avits@netvision.net.il

J. G. Schenker et al. (eds.), *Hot Topics in Human Reproduction*, Reproductive Medicine for Clinicians 3, https://doi.org/10.1007/978-3-031-24903-7_11

vitrification as a revolution in oocyte cryopreservation, making it sufficiently effective and suitable for clinical use [4]. A 2010 landmark study demonstrated similar clinical birth rates after transfer of embryos derived from fresh and cryopreserved oocytes of young oocyte donors [5]. Oocyte vitrification has enabled both oocyte banking for oocyte donation, as well as fertility preservation before gonadotoxic treatments and other medical conditions or therapies which are potentially harmful for the ovaries. One unique application of oocyte vitrification is for women who wish to address age-related fertility decline. This concept was first referred to as "social" freezing and later to "elective" freezing, but more recently termed "planned" [6]. The rationale behind this application is that fertility declines with age due to reduced oocyte number and quality, while the uterus retains its potential to bear a pregnancy. Therefore, oocytes cryopreserved at a younger reproductive age may enable higher chances for pregnancy and birth than a woman's current oocytes at more advanced age.

Planned oocyte cryopreservation appears to be a reasonable option for women in their 30 s who are not currently interested in having children but who wish to maintain the potential possibility for future pregnancy using their own oocytes. This concept is now approved by fertility societies. In 2012, ESHRE approved social oocyte cryopreservation [7]. ASRM declared medical oocyte cryopreservation as no longer an experimental procedure, however advised against social oocyte cryopreservation in 2012 [8]. They later endorsed social oocyte cryopreservation in 2018 [6]. Indeed, a major rise in oocyte cryopreservation has been noted in the past decade [9]. This appears to be mostly related to an increase in POC rather than for other indications, such as prior to medical gonadotoxic treatments [10].

Present Situation

Women who choose to undergo POC have higher than average education and socioeconomic status. Widespread public attention was drawn to the option of POC when leading technology companies offered to compensate employees for oocyte cryopreservation, possibly presenting POC as a mainstream practice for sophisticated "career women" [11]. However, the main reason reported by women who underwent POC was lack of a partner for parenthood [12–14]. Of note, POC has also become an option among conservative women, including women who had no previous sexual experience [14]. POC has mainly been performed by women in their 30 s, mean age 36–39. Approximately 15% of women underwent POC at age 40 and above [12, 14, 15]. The mean number of oocytes cryopreserved per cycle is around 10 according to most studies, but the standard deviation is high. Of note, in one study, 20% of women had a low ovarian response, defined as three or fewer oocytes suitable for cryopreservation [16].

Women who underwent POC have tended not to utilize their oocytes in the short term. Follow-up studies of women who underwent POC indicate that 20–48% tried to conceive naturally or by ART, in most cases without using their cryopreserved oocytes [12, 17–19]. The explanation for the low utilization rate of cryopreserved oocytes is that most single women (who initially chose POC to allow for the option of future pregnancy with a partner) still prefer not to pursue single parenthood by choice [14]. It seems that women undergo POC to "buy time," but eventually some women reevaluate their options and choose elective single parenthood. This again demonstrates how POC preserves a women's autonomy to make reproductive choices. Another possible rationale for this choice is the wish to maintain the option of an additional pregnancy at an older age, thus first attempting to conceive naturally while still possible. Longer follow-up periods are required to explore the full potential of cryopreserved oocytes.

Outcome

POC has been performed in many countries in large numbers since about 2010. Reported utilization rates to date are about 15%, with data on clinical outcomes limited by small sample size [12, 14, 17–22]. There is only one large study reported in the literature [15]. The mean age of women returning to use their cryopreserved oocytes is 36–9, with delivery rates of 26–34% per woman (Table 11.1). In the only study reporting outcomes of women who underwent POC at age 35 and younger, almost 70% had a live birth [15]. The reported birth rate was lower than 10% in women who had POC after age 40 [14, 20, 22, 23]. According to the largest study to date, the number of cryopreserved oocytes was strongly related to birth rates. In younger women, birth rates for women who preserved ten oocytes were 43%; this rose to 78% for women who preserved 20 oocytes. Above age 35, birth rates for 10 oocytes was 25% and for 20 oocytes 50% [15]. Since the mean number of oocytes

Table 11.1 Summary of reports on outcome of planned oocyte cryopreservation ($n > 10$)

Author, year	Women using cryopreserved oocytes	Age at cryopreservation	Number oocytes cryopreserved per woman	Post-thaw oocyte survival rate (%)	Live birth per women (%)
Cobo et al. (2018) [15]	641	37.2	9.8	91 ≤ age 35 82 > age 35	69 ≤ age 35 26 > age 35
Wennberg et al. (2019) [22]	38	38.7	12.8	78	26
Gurtin et al. (2019) [21]	30	37.7	NA	NA	27
Blakemore et al. (2021) [20]	80	38.2	14.3*	74	34
Leung et al. (2021) [23]	68	38 ± 2	14.4 ± 7.9	85	32

*Number of oocytes thawed per patient

at 37–8 is approximately 10, for most women, more than one cryopreservation cycle is advised in order to maximize the full potential of POC.

Emotional Aspects of POC

The high prevalence of psychological stress, anxiety, and depressive symptoms among women undergoing ART is well recognized [24]. Given that women who elect POC do not suffer from infertility, but rather make an informed decision to choose an elective procedure, it would be reasonable to assume less stress than that associated with conventional ART. However, researchers reported that women who underwent POC have similar responses quality-of-life questions compared with conventional IVF patients [25]. They perceived the experience of oocyte cryopreservation as difficult, emotionally more than physically [14]. This may in part be a reflection of the unsatisfactory relationship status that led the woman to elect POC, rather than the actual experience of the process itself. Research revealed varied emotional reflections, including women who reported positive emotional experiences including empowerment, satisfaction, and hope, while others described negative feelings such as sadness and depression and "going at it alone" [26]. Many women described a complicated and multi-faceted experience, including both positive and negative feelings [14]. Clearly, this is a unique group of patients in reproductive medicine, and specific tailored consideration of their emotional needs is warranted. Despite the perceived difficulties, most women who underwent POC reported high rates of satisfaction regarding their decision to undergo POC, and most did not express regret.

Ethical and Societal Aspects of Planned Oocyte Cryopreservation

Prior to the advent of effective oocyte cryopreservation, IVF technology was offered only to women or couples with infertility, or the imminent threat of infertility, such as prior to initiation of gonadotoxic chemotherapy. POC was termed a revolution in the field of reproductive medicine, as for the first time, a medical treatment was offered for those with no diagnosis of infertility, but to women at potential risk of fertility decline.

Media coverage led to extensive awareness regarding POC, which was described as granting "emancipation" for women. Social and ethical aspects of POC sparked much debate and discussion, including the preferred terminology for this novel concept.

The initial ethical terminology included autonomy, empowerment, and enabling informed choices. These terms, combined with the earlier nomenclature of "elective" oocyte cryopreservation, ignoe the reality that women were not usually actively choosing to be single and to limit their reproductive choices [13]. However, most of the women requesting oocyte cryopreservation were single by default and actually would have rather "elected" to be in a stable relationship and a position to start a family without medical intervention.

The term "AGE—banking" or oocyte cryopreservation for Anticipate Gamete Exhaustion was promoted as a prudent preventative measure, implying a looming medical threat necessitating timely intervention [27]. Other centers described "circumventing age-related fertility decline," creating the fallacy that women could control or override their natural fertility, which in turn served to further justify the widespread popularity and access to this previously exclusively medical technology.

POC, as a new technology, raises several unique ethical considerations which require further contemplation.

Marketing and Patient's Expectations

During this "internet age," POC is an example of "direct to consumer marketing" of a medical procedure. It has been posited that this aspect of reproductive medicine has developed from a medical discipline to an industry. Anxious women are guaranteed confidence for future genetic parenthood by undergoing a simple medical procedure.

Studies have portrayed fertility clinic websites advertising POC persuasively rather than informatively, emphasizing benefits while remaining deliberately less transparent about chances of success [28, 29]. This preys on the concern and anxiety common to this population and may contribute to unrealistic expectations, as well as influencing women to invest time, money, and confidence in POC. Women may simultaneously lose the opportunity to conceive using their own oocytes when still possible, by focusing on POC as the panacea.

The proposed "insurance policy" to enable women to "freeze" their fertility may give a false sense of security and hope. However, pregnancy and birth are not guaranteed. Extrapolating the current data reported in the literature indicates that according to age and number of oocytes cryopreserved, many, if not most, women who undergo POC may not achieve a live birth. Moreover, POC is expensive and not without risk. Hormonal stimulation may have emotional and physical side effects. Although the risk of ovarian hyperstimulation syndrome is essentially eliminated with the use of gonadotropin-releasing agonist for ovulation triggering with GnRH antagonist cycles [30], oocyte retrieval is associated with potential risk of anesthesia, bleeding, and infection. For some women, oocyte retrieval via the conventional vaginal approach is not feasible for religious reasons, and thus abdominal retrieval is necessitated. This can be more painful and possibly entail more risk than vaginal collection.

Unlike conventional ART, when a specific cycle's outcome is readily known, the success rates of POC remain theoretical until the woman returns to thaw her oocytes. At present, accurate, personal prediction for success of POC is challenging since actual results are somewhat limited. Many clinicians rely on predictive or extrapolatory models, rather than concrete data; little is available specifying age and number of oocytes cryopreserved. Therefore, caution is required when counseling women regarding predicted birth rates, in order to avoid unrealistic expectations for success. Appropriate counseling should also consider alternatives, such as fertilization of oocytes with donor sperm before cryopreservation, or fertility treatment/insemination with donor sperm.

Funding

POC is currently not funded by public health systems. Women—usually single—pay considerable amounts out of pocket. Public funding or subsidy of POC has been the subject of much controversy. Similar to debate regarding public IVF, POC has been promoted as a choice that should be independent of socioeconomic status and income. Various economic and statistical models have been presented to demonstrate that the costs of POC at a younger age may be far less expensive than future inefficient fertility treatments to achieve pregnancy at advanced maternal age [31]. Availability and accessibility of POC may also serve to increase awareness of age-related fertility decline which is generally underestimated by lay people. Possibly, setting a maximal age limit for public funding of POC, for example, 35, may encourage women to consider this option at an earlier age, making it more efficient in terms of future fertility.

There is also criticism against public funding of POC, as public resources are scarce and limited. First, age-related fertility decline is a "natural", expected process, rather than a disease state as other etiologies for infertility. This is exemplified by the term "non-medical" as suggested by some to describe POC, in contrast to "medical" egg freezing before cancer treatments, for example. Therefore, the financial cost of a woman's personal choice to postpone birth to an older age should not be subsidized by the state or national health budget. Another argument against public funding is the relatively low efficiency of POC, at least as performed today, i.e., by women at their late 30 s, and present low utilization rates. In these circumstances, POC is unlikely to be cost efficient.

Long-term follow-up studies on women who underwent POC consistently cite the leading cause for delaying pregnancy was not wanting to have a child without a partner, rather than career or financial issues [13, 14]. These women usually are aware of age-related fertility decline and would have started a family if they could. The fact that infertility naturally declines and ends with age is not a reason to withhold funding from medical treatment.

Low utilization rates may be considered a limitation in terms of public funding of fertility preservation. One could argue that utilization of sperm cryopreserved prior to chemotherapy is also low; however, this is routinely recommended. In addition, utilization rate and efficacy are also quite small for female fertility preservation procedures for women with cancer. However, such procedures are acceptable today and funded in some countries.

The important distinction lies in the definition of POC as a truly preventative technology, with a possibly curative future function. Countries such as France, which currently perceive POC as part of the medical fertility preservation treatment options, are reconsidering their budget allocation and funding [32].

POC as a Work Perk

POC as a potential advantage which empowers career women became the subject of much debate when Apple and Facebook offered free egg freezing to their female employees in 2014. This was later adopted by several high-tech, investment, and business employers as a "perk" for their female employers. Medical technology was presented as a ticket for women to be able to preserve equal rank with their male counterparts and continue to progress rising up the corporate ladder, without needing to slow down, or worse drop out, to have children. Funded POC claimed to provide these women with a pro feminist alternative to allow them to overcome the potential limits of their reproductive reality. This was lauded by some as a revolutionary option to enable women to enjoy more flexibility and experience less pressure regarding their declining ovarian reserve.

Conversely, the offer of free POC may be considered to actually be limiting the reproductive choices of the women it claims to liberalize. Women may choose, but they should not or must not choose to start a family during the critical years of their career. Promotion of POC may exert pressure on women to delay motherhood and rather turn to fertility preservation. Funded POC comes with the expectations that women will choose this option; women who may have been considering motherhood may feel that dedication and commitment to their career demands their choice of POC.

Birth at Advanced Age

The utilization of cryopreserved oocytes may enable childbearing at advanced chronological age, even beyond 60. While many countries have clear age limits for women requesting egg donation, established by legal and healthcare systems, the future thaw and use of oocytes preserved by POC may be perceived as somewhat different as they are the woman's OWN oocytes. This possible situation may demand revision of regulations and legislation.

Non-utilized Oocytes

At present, utilization rates of cryopreserved oocytes are low. In the longest follow-up study to date, 38% returned to use their oocytes after 10–15 years [22]. Regardless of the reason for the low return rate, a significant portion of cryopreserved oocytes will likely remain cryopreserved. This situation may create further technical and logistic strain on IVF labs, costs for patients, and a potential ethical or religious predicament regarding the disposal of these oocytes. A possible solution may be establishing a time limit on the period of cryopreservation, such as 10 and later 12 years instituted in the UK. However, this policy attracted much criticism, with concern that women may feel pressured to thaw and fertilized their oocytes before they would naturally choose to do so or forced to discard them. This may also result

in women choosing to cryopreserve oocytes at an older age than optimal, to prevent the potential pressure 10 years after POC.
Another potential solution is reassignment of unused oocytes for women seeking oocyte donation. For some women who are unable to choose disposal of their oocytes, they may be able to more readily accept knowing they may benefit a woman or couple in need.

Summary and Recommendations

How should we counsel women considering POC? For a woman in her mid-30 s and beyond, obviously the best chances of achieving a live birth with one's own oocytes is attempting to conceive promptly. When pregnancy is not a feasible option, for personal or other reasons, women considering POC may benefit from learning concrete outcomes emerging from studies reporting expected efficacy, rather than non-validated prediction models. It appears that approximately 1 in 3–4 women having POC at age 37–8 will achieve a birth, depending on number of oocytes cryopreserved [25]. These chances sharply decline after age 40. Women considering POC should be advised to complete the process before age 35 for optimal outcome. For most women, additional cycles are advised in order to accrue more oocytes, which is significantly associated with increased chances for live birth.

Many women are less proactive and delay decisions regarding pregnancy until their 40 s. According to the large UK database, ART cumulative delivery rates at age 40–42 are 31% per women [33]. But according to single center reports, this figure declined to 20% age 43 [34] and 4% at 44–5 [35]. Therefore, although disappointing for those searching for an "insurance policy" for future fertility, POC even at age 37–8 offers a better chance than having fertility treatments using a woman's own oocytes beyond age 42. Clearly, the key to consultation is sensitively providing comprehensive, accurate, and concise information regarding each option.

POC may be a promising option for women who are at risk of infertility at advanced reproductive age, especially if performed at an optimal age. It may enable women to conceive genetically related children at advanced reproductive age and reduce futile ART treatments and the consequent emotional distress and disappointment. Detailed information regarding efficacy of POC, mostly for young woman, is vital and will improve patient consultation and enable women to realize appropriate and personalized choices.

References

1. Ferraretti AP, Nygren K, Andersen AN, Mouzon J, Kupka M, Calhaz-Jorge C, et al. Trends over 15 years in ART in Europe: an analysis of 6 million cycles. Hum Reprod Open. 2017;2017:hox012.
2. Iussig B, Maggiulli R, Fabozzi G, Bertelle S, Vaiarelli A, Cimadomo D, et al. A brief history of oocyte cryopreservation: arguments and facts. Acta Obstet Gynecol Scand. 2019;98:550–8.

3. Kuleshova L, Gianaroli L, Magli C, Ferraretti A, Trounson A. Birth following vitrification of a small number of human oocytes: case report. Hum Reprod. 1999;14:3077–9.
4. Kuwayama M, Vajta G, Kato O, Leibo SP. Highly efficient vitrification method for cryopreservation of human oocytes. Reprod Biomed Online. 2005;11:300–8.
5. Cobo A, Meseguer M, Remohí J, Pellicer A. Use of cryo-banked oocytes in an ovum donation programme: a prospective, randomized, controlled, clinical trial. Hum Reprod. 2010;25:2239–46.
6. Ethics Committee of the American Society for Reproductive Medicine. Planned oocyte cryopreservation for women seeking to preserve future reproductive potential: an ethics committee opinion. Fertil Steril. 2018;110:1022–8.
7. Dondorp W, Wert G, Pennings G, Shenfield F, Devroey P, Tarlatzis B, et al. Oocyte cryopreservation for age-related fertility loss. Hum Reprod. 2012;27:1231–7.
8. American Society for Reproductive Medicine. Mature oocyte cryopreservation: a guideline. Fertil Steril. 2013;99:37–43.
9. Johnston M, Richings NM, Leung A, Sakkas D, Catt S. A major increase in oocyte cryopreservation cycles in the USA, Australia and New Zealand since 2010 is highlighted by younger women but a need for standardized data collection. Hum Reprod. 2021;36:624–35.
10. Human Fertilization and Embryology Authority. Egg freezing in fertility treatment trends and figures 2010–2016 final. 2018. https://www.hfea.gov.uk/media/2656/egg-freezing-in-fertility-treatment-trends-and-figures-2010-2016-final.pdf.
11. Tran M. Apple and Facebook offer to freeze eggs for female employees. The Guardian. 2014. https://www.theguardian.com/technology/2014/oct/15/apple-facebook-offer-freeze-eggs-female-employees.
12. Hammarberg K, Kirkman M, Pritchard N, Hickey M, Peate M, McBain J, et al. Reproductive experiences of women who cryopreserved oocytes for non-medical reasons. Hum Reprod. 2017;32:575–81.
13. Inhorn MC, Birenbaum-Carmeli D, Birger J, Westphal LM, Doyle J, Gleicher N, et al. Elective egg freezing and its underlying socio-demography: a binational analysis with global implications. Reprod Biol Endocrinol. 2018;16:70.
14. Tsafrir A, Holzer H, Miron-Shatz T, Eldar-Geva T, Gal M, Ben-Ami I, et al. "Why have women not returned to use their frozen oocytes?" A 5-year follow-up of women after planned oocyte cryopreservation. Reprod Biomed Online. 2021;43:1137–45.
15. Cobo A, García-Velasco J, Domingo J, Pellicer A, Remohí J. Elective and onco fertility preservation: factors related to IVF outcomes. Hum Reprod. 2018;33:2222–31.
16. Tsafrir A, Haimov-Kochman R, Margalioth EJ, Eldar-Geva T, Gal M, Bdolah Y, et al. Ovarian stimulation for oocyte cryopreservation for prevention of age-related fertility loss: one in five is a low responder. Gynecol Endocrinol. 2015;31:779–82.
17. Hodes-Wertz B, Druckenmiller S, Smith M, Noyes N. What do reproductive-age women who undergo oocyte cryopreservation think about the process as a means to preserve fertility? Fertil Steril. 2013;100:1343–9.
18. Balkenende EM, Dahhan T, van der Veen F, Repping S, Goddijn M. Reproductive outcomes after oocyte banking for fertility preservation. Reprod Biomed Online. 2018;37:425–33.
19. Wafi A, Nekkebroeck J, Blockeel C, Munck N, Tournaye H, Vos M. A follow-up survey on the reproductive intentions and experiences of women undergoing planned oocyte cryopreservation. Reprod Biomed Online. 2020;40:207–14.
20. Blakemore JK, Grifo JA, DeVore SM, Hodes-Wertz B, Berkeley AS. Planned oocyte cryopreservation-10–15-year follow-up: return rates and cycle outcomes. Fertil Steril. 2021;115:1511–20.
21. Gürtin ZB, Morgan L, O'Rourke D, Wang J, Ahuja K. For whom the egg thaws: insights from an analysis of 10 years of frozen egg thaw data from two UK clinics, 2008–2017. J Assist Reprod Genet. 2019;36:1069–80.
22. Wennberg A-L, Schildauer K, Brännström M. Elective oocyte freezing for nonmedical reasons: a 6-year report on utilization and in vitro fertilization results from a Swedish center. Acta Obstet Gynecol Scand. 2019;98:1429–34.

23. Leung AQ, Baker K, Vaughan D, Shah JS, Korkidakis A, Ryley DA, et al. Clinical outcomes and utilization from over a decade of planned oocyte cryopreservation. Reprod Biomed Online. 2021;43:671–9.
24. Boivin J. How does stress, depression and anxiety affect patients undergoing treatment? Curr Opin Obstet Gynecol. 2019;31:195–9.
25. Lee SS, Sutter M, Lee S, Schiffman MR, Kramer YG, McCulloh DH, et al. Self-reported quality of life scales in women undergoing oocyte freezing versus in vitro fertilization. J Assist Reprod Genet. 2020;37:2419–25.
26. Baldwin K, Culley L. Women's experience of social egg freezing: perceptions of success, risks, and "going it alone". Hum Fertil. 2018;23(3):186–92.
27. Stoop D, van der Veen F, Deneyer M, Nekkebroeck J, Tournaye H. Oocyte banking for anticipated gamete exhaustion (AGE) is a preventive intervention, neither social nor nonmedical. Reprod Biomed Online. 2014;28:548–51.
28. Avraham S, Machtinger R, Cahan T, Sokolov A, Racowsky C, Seidman DS. What is the quality of information on social oocyte cryopreservation provided by websites of Society for Assisted Reproductive Technology member fertility clinics? Fertil Steril. 2014;101:222–6.
29. Gürtin ZB, Tiemann E. The marketing of elective egg freezing: a content, cost and quality analysis of UK fertility clinic websites. Reprod Biomed Soc Online. 2021;12:56–68.
30. Humaidan P, Kol S, Papanikolaou EG, Copenhagen GnRH Agonist Triggering Workshop Group. GnRH agonist for triggering of final oocyte maturation: time for a change of practice? vol. 17. Hum Reprod Update; 2011. p. 510–24.
31. van Loendersloot LL, Moolenaar LM, Mol BWJ, Repping S, van der Veen F, Goddijn M. Expanding reproductive lifespan: a cost-effectiveness study on oocyte freezing. Hum Reprod. 2011;26:3054–60.
32. De Proost M, Johnston M. The revision of the French bioethics law and the questions it raises for the future of funding for egg freezing. Reprod Biomed Online. 2021;44(4):591–3. https://doi.org/10.1016/j.rbmo.2021.12.002.
33. Smith ADAC, Tilling K, Nelson SM, Lawlor DA. Live-birth rate associated with repeat in vitro fertilization treatment cycles. JAMA. 2015;314:2654–62.
34. Lebovitz O, Haas J, James KE, Seidman DS, Orvieto R, Hourvitz A. The expected cumulative incidence of live birth for patients starting IVF treatment at age 41 years or older. Reprod Biomed Online. 2018;37:533–41.
35. Raz N, Shalev A, Horowitz E, Weissman A, Mizrachi Y, Ganer Herman H, et al. Cumulative pregnancy and live birth rates through assisted reproduction in women 44–45 years of age: is there any hope? J Assist Reprod Genet. 2018;35:441–7.

12 Medical, Social, Legal, and Religious Aspects of Genetic Donation

Yoel Shufaro, Alyssa Hochberg, and Joseph G. Schenker

Introduction

Sperm, oocyte, and embryo donation are an integral part of the management of infertility when the existing technologies cannot resolve the basic biological problem. Donor insemination is the oldest modern treatment for male infertility and was introduced at the beginning of the twentieth century. Oocyte donation became available once ovarian stimulation became an integral part of assisted reproduction, resulting in surplus oocytes which can be donated to women of advanced age or with an inadequate ovarian reserve. These reproductive options, separately or joined in the form of combined gamete or embryo donations, are a remedy for childlessness, especially in an era of rising maternal age and single parenting. These donations are also significant in the aspect of genetic material donation, which the medical profession and society should consider not only from the interest of the infertile women or couple but also examining the interests of the offspring.

Oocyte Donation

Women of advanced age can conceive and deliver following the transfer of embryos originating from young donor oocytes. Case reports of deliveries in women well over 60 have been reported in peer-reviewed literature and also in popular media [1–4]. Since aging of the uterus is slower than that of the ovaries, the successful

Y. Shufaro (✉) · A. Hochberg
Infertility and IVF Unit, Beilinson Hospital, Rabin Medical Center, Petah Tikva, Israel

Sackler Faculty of Medicine, Tel-Aviv University, Tel Aviv, Israel
e-mail: yoelshufaro@tauex.tau.ac.il

J. G. Schenker
The Hadassah-Hebrew University Medical Center, Jerusalem, Israel

J. G. Schenker et al. (eds.), *Hot Topics in Human Reproduction*, Reproductive Medicine for Clinicians 3, https://doi.org/10.1007/978-3-031-24903-7_12

implantation of embryos from young donor oocytes into the uteri of perimenopausal and menopausal women is readily accomplished [5, 6]. Embryo implantation is scarcely affected by the endometrial age, and therefore oocyte donation in women of advanced age is as successful in achieving pregnancies and live births as is assisted reproduction in the donors' age group [7]. The uterus retains its receptivity to embryo implantation for a substantial period of time after the ovarian germ cell reserve diminishes, as long as adequate endogenous or exogenous hormonal support exists or is provided. As a consequence, women of very advanced age, even up to the seventh decade of life, are able to conceive, carry pregnancies, and deliver live-born babies to whom they are not genetically related. With the increase in maternal age in developed countries, the number of women contemplating and achieving pregnancies of non-genetic offspring at an age previously considered adequate for grandparenthood is constantly rising.

The medically and ethically adequate availability of oocytes from young donors is the basis for peri- and post-menopausal pregnancies and deliveries. Donor oocytes are used in cases of advanced maternal age and premature ovarian insufficiency, low ovarian reserve, disorders inherited through the mitochondrial DNA, or other maternally inherited disorders in which pre-gestational testing is not feasible. Donor oocytes were historically obtained initially from IVF patients donating surplus oocytes, but currently, most donor oocytes originate from financially compensated volunteers who are termed "donors" despite being paid. The embryos originating from donor oocytes can be transferred fresh if the recipient is synchronized with the donor or cryopreserved and transferred in a different cycle [8]. The latter alternative also facilitates cross border traffic of donor oocytes, fertilized or unfertilized, from countries in which the compensated donors live to the recipients' place of residence.

In menopausal recipients, artificial endometrial preparation with exogenous estrogens followed by addition of progesterone, in a manner similar to artificial cycle for frozen-thawed embryo transfer, is required [5]. After pregnancy is confirmed, the endometrial support administration should be continued until placental autonomy occurs. In cases where ovarian activity still exists, embryos can be transferred based on the endogenous ovarian cycle. The average reported global ongoing pregnancy and delivery rate is currently approximately 50% per transfer [9].

The oocytes are aspirated from paid volunteers compensated for their "expenses." Candidates must be younger than 35, healthy and free of contaminants transmitted through body fluids, such as hepatitis viruses, HIV, or syphilis. They are screened for autosomal and X-linked hereditary disorders, as well as for structural chromosomal rearrangements. Broader screening of recessive traits, including several hundred inherited disorders, is currently under endorsement by quite a few oocyte donation programs in order to minimize the genetic risk associated with this procedure. Those who volunteer to donate oocytes for others undergo moderate ovarian stimulation, because excessive stimulation, in addition to being unsafe for the donor, is also detrimental to the quality of the oocytes. The standard stimulation protocol for oocyte donors is the antagonist protocol with a GnRH agonist, and not hCG, used for triggering oocyte maturation [10]. This approach has a high oocyte yield and maturity rate, in addition to effectively protecting the donor from early ovarian hyper-stimulation syndrome. The implantation, clinical pregnancy, and live birth rates are unaffected when embryos from oocytes obtained with a GnRH agonist trigger are transferred to the uteri of separately prepared recipients.

Medical Complications in Recipients of Donor Oocytes

In countries in which oocyte donation from designated compensated volunteers is performed, relevant legislation or directives exist, set to protect the health, rights, and anonymity of the volunteers [11] and minimize the health hazards to which they are exposed. On the other hand, there is a relative paucity of regulations, other than a maximal age limit, concerning the management of the recipients, most of them women of advanced age. The medical pregestational evaluation required for ascertaining the suitability of these women for pregnancy and delivery is not rigorously defined despite the significant hazards that pregnancy might impose on them. Thus, while the use of oocytes from young (paid) volunteers reduces the fetal genetic hereditary and non-hereditary risk, it is quite well-established that pregnancy in the older population potentially constitutes a major maternal-fetal health risk during pregnancy.

The respiratory, hemodynamic, renal, and endocrinologic changes in pregnancy are a stressful event in young women. The cardiac output gradually increases to 140% of the baseline value and transiently even more than that during labor. At advanced age, the adaptation capacity of the cardiovascular system might be impaired, even in apparently healthy patients. The pulmonary respiratory volumes and effort, as well as the renal glomerular filtration rate, rise significantly during pregnancy [12]. Additionally, occult hypertension, heart diseases, diabetes, chronic lung diseases, renal diseases, and other conditions might exist at an advanced age [13, 14], jeopardizing the health of the mother and fetus, up to the point necessitating premature termination of the pregnancy. Age is also a risk factor for the occurrence of gestational trophoblastic diseases, fibroids, and urinary tract infections that might complicate pregnancies as well [15–17].

Setting up an age limit for conception attempts and determining the medical evaluation required for the candidates in order to go through pregnancy safely is under constant discussion.

The prevalence of pregnancy-induced or exacerbated hypertension, preeclampsia, impaired glucose tolerance, and frank diabetes are all in correlation with age [13, 18]. Moreover, in women over 50, the occurrence of these conditions is even more elevated in comparison to the 40–49 years age group [14]. These observations have been confirmed by other studies, reaching a peak of 63% for risk for any complication requiring hospitalization [18–20]. In singleton live births of women over 45, the risk for preeclampsia in oocyte donation recipients was 12.6% compared to 1.1% in spontaneous pregnancies at the same age. In contrast to natural conceptions, the preeclampsia risk in oocyte donation recipients over 45 was constant and was unaffected by previous parity [21].

Age is also an independent risk factor for placental abruption and malpresentation (placenta previa) [16, 18], probably as a result of uterine microvasculature changes [22]. In the case of multi-fetal gestations—a common result of assisted reproduction—the prevalence of hypertensive and placental complications is even higher [19]. Additionally, the likelihood of preterm labor or complications necessitating premature delivery is also increased with maternal aging [2, 14, 18–20, 23, 24]. The risk of preterm deliveries and low birth weight is significantly increased over age 50 for all types of gestations—multiples and singletons [19]. Therefore, taking into consideration the high success rates of assisted reproduction with young donor oocytes on the one hand, and the very high-risk that multifetal gestation bears at an advanced age on the other, a mandatory single embryo transfer policy is strongly recommended under these circumstances.

Direct and indirect maternal mortality also correlate with age. In developed countries, the primary reasons for such tragedies are mainly exacerbations of pre-existing medical conditions or the occurrence of dramatic severe preeclampsia, placental abruption, postpartum hemorrhage, and thromboembolic events [25]. The almost universal performance of cesarean sections for delivering women of advanced age who conceived through oocyte donation does not contribute to the maternal mortality [2]. Sporadic maternal deaths of oocyte recipients of advanced age were reported [25], but underreporting of such cases can be assumed. Nevertheless, even at advanced age, maternal mortality is still a rare event in developed countries in which up-to-date prenatal care is available. Although the relative risk of maternal mortality at advanced age is increased, the absolute risk with proper screening and adequate antenatal care is still very low. After adequate screening of a healthy population, the maternal morbidity and mortality is low enough not to ban oocyte donation and pregnancy at an advanced maternal age [26]. With proper maternal-fetal antenatal care, both the maternal and neonatal outcomes are reasonably good.

The Neonate at Advanced Maternal/Parental Age

While the use of oocytes from young donors reduces the fetal aneuploidy and malformation risk expected in the advanced maternal age group [7], the prevalence of obstetrical complications such as low birth weight, prematurity, and stillbirths is increased in neonates of mothers of advanced age [2, 7, 14, 23, 27, 28]. This is the result of the increased prevalence of complications necessitating pre-term delivery, such as pre-term labor and abnormal placental function. On the other hand, the prevalence of low Apgar scores, neonatal asphyxia, and metabolic acidosis is not increased in comparison to younger women [18, 23].

The long-term psychological and social impact of being the child of an elderly mother, father, or parents varies greatly between countries, populations, and societies, in accordance with culture, social norms, life expectancy, and quality of life at the advanced age. Opponents to pregnancies in older women or parents reason this position based on the interest and welfare of the future offspring, thus implying that older individuals might be or are less suitable parents [29], based on a greater generation gap, growing up without grandparents, or parental age-associated medical morbidities and a shorter life expectancy [30]. On the other hand, older people are more mature and experienced than younger ones and have more free time, as well as emotional and material resources to nurture children. The deep-seated desire for an offspring might be of more benefit than harm to the child [31]. Taking all these into consideration, it is reasonable to assume that in societies with a longer healthy life expectancy, advanced parental age has little if any negative impact on the offspring.

Sperm Donation

IVF and intracytoplasmatic sperm injection (ICSI), especially in combination with testicular sperm retrieval procedures, facilitate genetic parenthood for a vast majority of couples with male factor infertility [32]. Still, donor sperm is a treatment modality for severe male infertility, either as a first line or second line of treatment, when exploration for testicular sperm in azoospermic men failed or if the fertilizing capacity of the available partner spermatozoa is reduced. Sperm donations are also used by women without a male partner or with a trans male partner [33].

The most prevalent form of sperm donation is an anonymous donation through a sperm bank service, and limitations do exist in different countries concerning the number of recipients one sperm donor can donate to. Directed donations in which sperm is donated by an individual to a particular female recipient who is familiar (but not intimate) with him, without bearing paternal duties, are accepted in some countries. Sperm donors are compensated volunteers of the local legal consent age, who have a normal spermatogram. They are screened meticulously for physical and mental health conditions, genetic conditions, and carriership of occult infectious agents such as HIV, hepatitis B and C, and syphilis. Once a donor is enrolled, a 3–6-month quarantine period before the sperm can be used is warranted in order to ensure that repeated serology is negative, ruling out an infection that might have been present when the sperm sample was provided. Fresh sperm donations are currently outside the standard medical practice because the risk of transmitting infectious agents cannot be eliminated without a proper quarantine period [33].

Donated sperm can be used for artificial insemination (AID) or for IVF. The outcome of the treatment depends on its type, the patients age, and other patient data, not on the sperm used. In a meta-analysis of the clinical outcomes of sperm donation including eight studies, donor sperm neonates were not at increased risk of being born with low birth weight, preterm, or with increased incidences of birth defects, than were spontaneously conceived neonates [34].

Ethical and Legal Considerations

The principal ethical considerations regarding gamete donation are protection of the donors' (male and female), recipients', and offspring privacy, as well as the medical safety of the oocyte donors and recipients. While the latter are mainly medical issues previously discussed, the privacy dilemma is a major ethical consideration. The other ethical and legal dilemma is the financial compensation of those who volunteer to donate their gametes and the complicated issue of embryo donation.

Gamete donors can be either anonymous or known to the couple, partially or fully [33]. In most countries the anonymity of the donors is preserved. Traditionally, sperm donation was mostly clandestine and was not accepted as a social and marital norm. Donor insemination was considered in some societies as being illegitimate or constituting adultery. Secrecy was also in the male partners' interest since it protected him from the social stigma of sterility, associated with male dysfunction in many cultures. Since female infertility is more acceptable and described in different, older scripts, cases of oocyte donation in which the conception results from assisted reproduction are more socially acceptable even in traditional, conservative societies. Nevertheless, in most countries both sperm and oocyte donor anonymity are protected by law for the following reasons:

1. Donors' concerns about legal and social parenthood liability if their identity is disclosed might preclude gamete donation altogether.
2. Protection of the future privacy of the donors and especially their future families.
3. Prevention of social embarrassment for the recipients.
4. Prevention of parental confusion among gamete donations' offspring.

On the other hand, several universal and particular arguments for identifying gamete donors were raised, especially for the potential psychological benefit of the offspring. The main ones are:

1. Truthful disclosure as a universal value that outweighs accidental discovery based on physical discrepancy or blood type mismatch.
2. The individual's universal basic right to explore and uncover his/her biological identity, ancestors, and origin.
3. Equality with the offspring of spontaneous gestations who are familiar with their biological parents.
4. The relevance of the donors' evolving medical history to the health of the offspring.
5. Prevention of accidental consanguinity in the next generation.

It is interesting that despite these arguments, even where the disclosure of donor identity is the legal norm, most children that were conceived by sperm donation are not interested in this disclosure [35]. The bottom line is that consideration for protecting the donors and parents overcomes any considerations for protecting the rights of future, but presently unborn, offspring. If the theoretical rights of the latter were the dominant consideration, then gamete donations would be reduced to scarce or even null numbers in many countries. Some programs offer a form of compromise between absolute anonymity and full disclosure in the form of partial disclosure. This includes a variety of details regarding the physical appearance and biography of the gamete donor, hobbies, personality traits, and even exposure of the recipients to childhood and recent photos of the donor. In this manner a feeling of acquaintance is achieved without revealing the donor's identity.

Payment for Donation of Genetic Material

Most international ethical committees are against financial compensation for individuals who volunteer to donate their gametes to others. On the other hand, financial compensation is a serious drive for gamete donation. Local regulations in different countries provide a solution to this dilemma by authorizing compensation for the "time and expenses" of the volunteer and not for the gametes [33, 36]. Obviously, such payment would be lower for sperm donation than for oocyte donation. Ideally, there are almost no donor expenses in donating sperm, so it should be donated altruistically, and payment should not be the main motivation for donation [37]. Unfortunately, this is not the case in most of the world, and sperm donors are compensated in a financially attractive manner. In most countries, donor sperm for IUI or IVF is not covered by the public health system and the cost of the sperm donation is paid by the patient.

Oocyte donation is associated with a substantial donor effort, time input, and risk in undergoing ovarian stimulation and oocyte pick up, so an adequate compensation is warranted.

The Voluntary Licensing Authority for Human in Vitro Fertilization and Embryology in the United Kingdom has decided to allow centers to offer free procedures in return for donated eggs. Some centers offer a free IVF cycle treatment as compensation for excess egg donation. The American Society for Reproductive Medicine guidelines state that the donor should be compensated for direct and indirect expenses associated with their participation, inconvenience, and time and, to some degree, for the risk and discomfort undertaken. Payment should not be predicated on the number of oocytes donated and should not be the primary incentive for the donation [33]. Nevertheless, despite this statement, in the United States, a sum as high USD 8000 is paid to oocyte donors [38], much higher than one would expect as compensation for expenses.

Embryo Donation

With the introduction of cryopreservation as a routine practice in IVF, there is an excess of stored surplus embryos up to the point of a cryostorage space crisis. Donation of these embryos to patients or couples in need is an appealing idea, but raises substantial ethical and legal problems:

1. Who is the owner of undesired human cryopreserved embryos?
2. Can human embryos be sold or bought just as gametes are?
3. Who controls the disposition of the stored embryos in case of death of both or one of the progenitors or in the case of divorce?

The answers to those questions were provided by several ethical committees and legislators regulating ART. The legal status of the human embryo in cryostorage is difficult to establish. A cryopreserved embryo is not considered a human being for the purpose of criminal law. On the other hand, the cryopreserved embryo is not property. If a dispute arises between the couple who provided the sperm and oocyte from which the embryo was formed about its disposition, the embryo will remain cryopreserved until a legal or judicial decision has been reached.

Documentation and Registration

There is a consensus among medical professionals that keeping accurate medical records is essential. Record keeping has always been an important part of both medical practice and of quality assurance. In cases of gamete donation, it is also crucial for follow-up of the parties involved. It raises particularly difficult ethical and legal questions with regard to medical confidentiality and family privacy. The right to privacy is a fundamental human right. In the context of medical information that is personal and intimate, the concern for respect for the privacy of the participants is paramount. Truth-telling and candidness are values to be respected in the communication between physician and patient, and in the case of gametes and pre-embryo donation, it may be considered in the relationship between the physician, the donor, and the recipient. Candidness with the family after the birth of a child as to the method of his conception, or later as to the identity of the donor, is of a different nature. Society's (or the state's) intervention in the privacy and intimacy of the familial relationship, in order to force a greater openness, could be an invasion of the freedom of procreation decision-making that extends beyond the legitimate concern for the quality of services and for proper follow-up of the offspring. Registration and regulations in different countries, where gametes and pre-embryo donation are practiced, take into consideration the nature of the information to be maintained about the parties involved in the gametes and pre-embryo donation program. Thus, a distinction has been drawn between non-identifying and identifying information. The non-identifying information includes:

(a) Detailed description of physical characteristics, ethnic origin, etc.
(b) Medical history and genetic background.
(c) Social characteristics: education, profession, habits, interests, etc.

When identifying information is required, it will include full names, addresses, dates, and places of birth, as well as the IDs of the parties involved.

The responsibility for the collection of information should lie with the physician performing each stage of the donation procedure. There are different opinions regarding the storage of information: Where should it be kept? Who should have access to it? What kind of information should be released to the parties involved in the program?

In most countries where genetic material donation is practiced, the records of identifying and non-identifying information are kept and maintained by the physicians or medical institutions according to the regulations of the particular country. In some countries, it was suggested that the identifying information of the parties involved should be stored in the Central Government Registry. The advantages of a central state registry are:

(a) The information can be safely kept for long periods.
(b) There is a protected central control on the release of information.
(c) A central computerized national register may provide control over the number of donations made by each donor.
(d) It is of importance to restrict to a minimum the personnel who have access to this information.

Identifying material may be released in extreme situations according to the legislation in a specific country. The legislation should not be retrospective on current or past participants in the program. The identifying information can be released only if the parties involved have given their consent to it prior to the procedure. Conflicts of interest may arise between the parties involved—sperm, ovum, and pre-embryo donors, offspring, and parents—regarding disclosure and access to information.

Religious Aspects of Genetic Material Donation

Roman Catholic Church

The issue of human reproduction was discussed in the Congregation for the Doctrine of the Faith in February 1987, signed by Cardinal Joseph Ratzinger, and approved by Pope John Paul II (Doctrine of the Faith, 1987). The key value in the instructions is respect for the dignity of the human being. Fertilization is allowed when it is the result of a conjugal act, that is, sexual intercourse between husband and wife. Consequently, the instruction prohibits IVF—embryo transfer, surrogate motherhood, and cryopreservation of embryos. It also rejects AID and IVF on the grounds that this involves a separation between "the goods and meanings of marriage." This position eliminates any use of donor semen for artificial insemination or for IVF. Furthermore, artificial fertilization of a woman who is unmarried or widowed, whoever the donor may be, cannot be morally justified. The practice of ovum and embryo donation is prohibited on the same basis as sperm donation.

Other Christian Churches

The Eastern Orthodox Church supports the medical and surgical treatment of infertility. IVF and other assisted reproductive technologies are not absolutely rejected. However, the Church opposes gamete donation, especially AID, on the grounds that it constitutes an adulterous act.

The Baptist, Methodist, Lutheran, Mormon, Presbyterian, Episcopalian, United Church of Christ, Christian Science, Jehovah's Witness, and Mennonite religions have liberal attitudes toward infertility treatments. All denominations except Christian Science accept IVF with the spouse's gametes and no embryo wastage [39]. Christian Science poses no objection to artificial insemination but opposes IVF because of the drugs and surgical procedures used. The aforementioned religions oppose IVF with donated gametes and the practice of surrogacy.

Islam

The procedure of IVF embryo transfer is acceptable, but it can be performed only if it involves the gametes of a husband and wife. A third party is not acceptable, whether in providing the egg, spermatozoon, embryo, or uterus. If a marriage has come to an end through divorce or death of the husband, artificial reproduction cannot be performed on the woman even by using spermatozoa from her late husband. Islamic law strictly condemns the practice of AID on the grounds that it is adulterous. According to the Muslim faith, for example, a Muslim man can marry a Jewish or Christian woman, as the religion of offspring is linked to the father.

Oocyte donation is not permitted in Islam, since it involves the intervention of a third party [40, 41]. Islamic law limits a man to the marriage of four wives simultaneously. Donation of oocytes between wives is not permitted. Donation of embryos, according to Islam, is prohibited. Frozen embryos are the property of the couple alone.

However, according to Fatwa from Ayatollah Hussein Khomeini in 1999, egg donation was approved only for the Shia sector. According to Iranian law, oocyte donation can be permissible under certain circumstances.

According to the Druze religion (a minority group of less than 1,500,000 persons living in the Middle East and originating from Islam), donation of oocytes can be permitted only between sisters.

Judaism

Therapeutic insemination with donor spermatozoa (AID) is accepted by a portion of the Jewish population in Israel and is unacceptable to most rabbinical authorities. Rabbis have been discussing the principles involving AID for many centuries. Their discussions are based on ancient sources in the Talmud and codes of Jewish law dating back to the fifth century that mention procreation without intercourse.

Jewish law prohibits AID for a varicty of reasons: resemblance to incest, lack of genealogy, and problems related to inheritance. In addition, donors are violating the severe prohibition against masturbation. Many rabbinical scholars consider a child conceived through AID as having the status of "mamzer" (bastard), which severely limits prospects of marriage and implies a severe social handicap. Some rabbinical authorities permit AID if the donor is not a Jew. Rabbi Moshe Feinstein ruled that with the husband's permission and in the case where the infertile couple is suffering considerably, one may permit donor insemination, but specifically with the sperm from a non-Jewish donor. This eliminates some of the legal complications related to the personal status of the offspring. If the donor is a gentile, the child is not a "mamzer," but if the child is a girl, she is forbidden to marry a Cohen (a person with temple priest ancestry). Another reason for preferring non-Jewish donor sperm is to prevent future accidental consanguinity among the offspring of anonymous donors.

Oocyte (from single women) and embryo donations are allowed in Judaism, and the main issue is whether the religious status of the offspring should be based on the oocyte donor or the recipient. Jewish law dictates maternal determination of the religious status of the child. For purposes of lineage, the woman receiving the egg, rather than the woman donating the egg, is the mother, although the latter is certainly the genetic parent. If the recipient is Jewish, then the child is considered Jewish.

Hinduism

Assisted reproductive technologies are acceptable in Hinduism because there is no single authority to accept or reject on behalf of the faith. The most important condition is that the oocyte and sperm are from a legally married couple. In practice, artificial insemination with donor sperm and oocyte or embryo donation are performed with an anonymous donor. It is preferable that the sperm donor be a close relative of the husband.

Buddhism

The Buddhist religion is practiced by about 500 million people, representing 7–8% of the world's population. The largest Buddhist populations reside in China, Thailand, and Japan.

Buddhism of all types in various countries is individualistic, and even their scripture is not rigid.

There is no central Buddhist authority to pronounce on religious positions. Marriage within Buddhism does not have the high priority that it has in monotheistic religions. Any technology that is used to achieve pregnancy is morally acceptable, and treatment can be given to the married as well as to the unmarried.

In China, sperm, oocyte, and embryo donation for research is controlled by governmental regulation. Sperm donation is completely anonymous; only donors between the ages of 22 and 44 years are eligible for selection; donor sperm cannot be provided to single women or same-sex couples; and each sperm donor can only impregnate up to five women via AID or IVF.

In Japan, anonymous sperm and oocyte donation is practiced. Commercial oocyte donation is not permitted.

References

1. Antinori S, Versaci C, Gholami GH, Panci C, Caffa B. Oocyte donation in menopausal women. Hum Reprod. 1993;8(9):1487–90.
2. Benshushan A, Schenker JG. Age limitation in human reproduction: is it justified? J Assist Reprod Genet. 1993;10(5):321–31.
3. Pantos K, Meimeti-Damianaki T, Vaxevanoglou T, Kapetanakis E. Oocyte donation in menopausal women aged over 40 years. Hum Reprod. 1993;8(3):488–91.
4. Paulson RJ, Sauer MV. Pregnancies in post-menopausal women. Oocyte donation to women of advanced reproductive age: 'how old is too old?'. Hum Reprod. 1994;9(4):571–2.
5. Navot D, Laufer N, Kopolovic J, Rabinowitz R, Birkenfeld A, Lewin A, et al. Artificially induced endometrial cycles and establishment of pregnancies in the absence of ovaries. N Engl J Med. 1986;314(13):806–11.
6. Sauer MV, Paulson RJ, Lobo RA. Pregnancy after age 50: application of oocyte donation to women after natural menopause. Lancet. 1993;341(8841):321–3.
7. Soderstrom-Anttila V. Pregnancy and child outcome after oocyte donation. Hum Reprod Update. 2001;7(1):28–32.
8. Cobo A, Bellver J, Domingo J, Perez S, Crespo J, Pellicer A, et al. New options in assisted reproduction technology: the Cryotop method of oocyte vitrification. Reprod Biomed Online. 2008;17(1):68–72.
9. Noyes N, Hampton BS, Berkeley A, Licciardi F, Grifo J, Krey L. Factors useful in predicting the success of oocyte donation: a 3-year retrospective analysis. Fertil Steril. 2001;76(1):92–7.
10. Humaidan P, Kol S, Papanikolaou E. GnRH agonist for triggering of final oocyte maturation: time for a change of practice? Hum Reprod Update. 2011;17(4):510–24.
11. Schenker JG. Sperm, oocyte, and pre-embryo donation. J Assist Reprod Genet. 1995;12(8):499–508.
12. Sibai BM, Frangieh A. Maternal adaptation to pregnancy. Curr Opin Obstet Gynecol. 1995;7(6):420–6.
13. Paulson RJ, Boostanfar R, Saadat P, Mor E, Tourgeman DE, Slater CC, et al. Pregnancy in the sixth decade of life: obstetric outcomes in women of advanced reproductive age. JAMA. 2002;288(18):2320–3.
14. Salihu HM, Shumpert MN, Slay M, Kirby RS, Alexander GR. Childbearing beyond maternal age 50 and fetal outcomes in the United States. Obstet Gynecol. 2003;102(5 Pt 1):1006–14.
15. Jequier AM, Winterton WR. Diagnostic problems of trophoblastic disease in women aged 50 or more. Obstet Gynecol. 1973;42(3):378–87.
16. Naeye RL. Maternal age, obstetric complications, and the outcome of pregnancy. Obstet Gynecol. 1983;61(2):210–6.
17. Tuck SM, Yudkin PL, Turnbull AC. Pregnancy outcome in elderly primigravidae with and without a history of infertility. Br J Obstet Gynaecol. 1988;95(3):230–7.
18. Yogev Y, Melamed N, Bardin R, Tenenbaum-Gavish K, Ben-Shitrit G, Ben-Haroush A. Pregnancy outcome at extremely advanced maternal age. Am J Obstet Gynecol. 2010;203(6):558.e1–7.
19. Simchen MJ, Shulman A, Wiser A, Zilberberg E, Schiff E. The aged uterus: multifetal pregnancy outcome after ovum donation in older women. Hum Reprod. 2009;24(10):2500–3.

20. Simchen MJ, Yinon Y, Moran O, Schiff E, Sivan E. Pregnancy outcome after age 50. Obstet Gynecol. 2006;108(5):1084–8.
21. Dior UP, Laufer N, Chill HH, Granovsky-Grisaru S, Yagel S, Yaffe H, et al. Increased incidence of preeclampsia in mothers of advanced age conceiving by oocyte donation. Arch Gynecol Obstet. 2018;297(5):1293–9.
22. Zalud I, Shaha S. Three-dimensional sonography of the placental and uterine spiral vasculature: influence of maternal age and parity. J Clin Ultrasound JCU. 2008;36(7):391–6.
23. Dulitzki M, Soriano D, Schiff E, Chetrit A, Mashiach S, Seidman DS. Effect of very advanced maternal age on pregnancy outcome and rate of cesarean delivery. Obstet Gynecol. 1998;92(6):935–9.
24. Sheffer-Mimouni G, Mashiach S, Dor J, Levran D, Seidman DS. Factors influencing the obstetric and perinatal outcome after oocyte donation. Hum Reprod. 2002;17(10):2636–40.
25. Bewley S, Wright JT. Maternal death associated with ovum donation twin pregnancy. Hum Reprod. 1991;6(6):898–9.
26. Edwards RG. Pregnancies are acceptable in post-menopausal women. Hum Reprod. 1993;8(10):1542–4.
27. Schenker JG. The therapeutic approach to infertility in cases of ovarian failure. Ann N Y Acad Sci. 1991;626:414–30.
28. Kenny LC, Lavender T, McNamee R, O'Neill SM, Mills T, Khashan AS. Advanced maternal age and adverse pregnancy outcome: evidence from a large contemporary cohort. PLoS One. 2013;8(2):e56583.
29. Goold I. Should older and postmenopausal women have access to assisted reproductive technology? Monash Bioeth Rev. 2005;24(1):27–46.
30. Ethics Committee of the American Society for Reproductive Medicine. Oocyte donation to postmenopausal women. Fertil Steril. 2004;82(Suppl 1):S254–5.
31. Kortman M, Macklon NS. Oocyte donation in postmenopausal women: medical and ethical considerations. Obstet Gynaecol Reprod Med. 2008;18(6):168–9.
32. Almekaty K, Abomelha S, Thum Y, Nicopoullos J, Bracewell-Milnes T, Yap T, et al. Reporting on longitudinal live birth rates and cumulative delivery rates are more realistic outcome measures than sperm retrieval rates in couples undergoing mTESE-ICSI. Hum Fertil (Camb). 2019;22(2):139–44.
33. Practice Committee of the American Society for Reproductive Medicine, The Practice Committee for the Society for Assisted Reproductive Technology. Guidance regarding gamete and embryo donation. Fertil Steril. 2021;115(6):1395–410.
34. Adams DH, Clark RA, Davies MJ, de Lacey S. A meta-analysis of sperm donation offspring health outcomes. J Dev Orig Health Dis. 2017;8(1):44–55.
35. Lampic C, Skoog Svanberg A, Gudmundsson J, Leandersson P, Solensten NG, Thurin-Kjellberg A, et al. National survey of donor-conceived individuals who requested information about their sperm donor-experiences from 17 years of identity releases in Sweden. Hum Reprod. 2022;37(3):510–21.
36. Johnson M. Payments to gamete donors: position of the human fertilisation and embryology authority. Hum Reprod. 1997;12(9):1839–42.
37. Goedeke S, Shepherd D, Rodino IS. Support for recognition and payment options for egg and sperm donation in New Zealand and Australia. Hum Reprod. 2020;35(1):117–29.
38. Zakarin Safier L, Gumer A, Kline M, Egli D, Sauer MV. Compensating human subjects providing oocytes for stem cell research: 9-year experience and outcomes. J Assist Reprod Genet. 2018;35(7):1219–25.
39. Dunstan GR. In-vitro fertilization: the ethics. Hum Reprod. 1986;1(1):41–4.
40. Schenker JG. FIGO committee for the study of ethical aspects of human reproduction. Ethical aspects in the management of the severely malformed fetus. International Federation of Gynecology and Obstetrics. Ginekol Pol. 1996;67(11):585–6.
41. Schenker JG, Shushan A. Ethical and legal aspects of assisted reproduction practice in Asia. Hum Reprod. 1996;11(4):908–11.

Informed Consent in Reproductive Therapy

13

Ofra G. Golan

Introduction

The concept of informed consent has been developed during the second half of the twentieth century. It started as a legal doctrine in the American jurisprudence [1], but its underlying idea and basic principles have been widely accepted by other legal systems during the years. However, informed consent should not be looked at just as a legal requirement; first and foremost, it is an ethical issue. As Justice Kirby, then president of the Australian Law Reform Commission wrote [2]:

The fundamental principle underlying consent is said to be a right of self-determination: the principle or value choice of autonomy of the person. The principle is not just a legal rule devised by one profession to harass another. It is an ethical principle which is simply reflected in legal rules because our law has been developed by judges sensitive to the practical application of generally held community ethical principles.

Doctors tend to be quite familiar with the legal aspects of the requirement of patients informed consent, much more so in societies and in specialties where they are more exposed to malpractice lawsuits. Still, discussing the ethical aspects of the concept of informed consent would shed light on the reasoning and justification of the legal demands. This should help the doctor follow these demands more willingly and naturally rather than simply obey the rules set by the law (which, from this state of mind, sometimes seem not to be acceptable or applicable). It may also give the doctor guidance where the law is not clear or is unknown to him/her (*).

*For convenience, from now on, the patient will be related to in this chapter in feminine gender and the doctor in masculine gender.

O. G. Golan (✉)
The Israel Academic College in Ramat Gan, Ramat Gan, Israel

J. G. Schenker et al. (eds.), *Hot Topics in Human Reproduction*, Reproductive Medicine for Clinicians 3, https://doi.org/10.1007/978-3-031-24903-7_13

What Is Meant by Informed Consent?

The phrase informed consent has two components:

1. Informed—which means based on adequate information and deliberation
2. Consent—which means a voluntary, uncoerced decision, made by a sufficiently competent or autonomous person, to accept rather than reject some proposed course of action that will affect him or her [3]

An informed consent is an autonomous authorization of a medical intervention or of involvement in research by individual persons [4].

The concept of informed consent can be defined as the idea that the patients' free consent, based on all relevant information necessary for the decision to undergo the treatment, is a prerequisite to any medical intervention (excluding exceptional circumstances).

This implies true communication between doctor and patient: disclosure of information from one party and its comprehension by the other.

Ethically valid consent is a process of shared decision-making based upon mutual respect and participation, not a ritual to be equated with reciting the contents of a form that details the risks of particular treatments [5]. Thus, informed consent is not just an act, or set of acts; it is a process. It is a process in which doctor is discussing the treatment options with his patient. The doctor suggests a certain intervention and explains his considerations for this course of action, mentioning the pros and cons of the suggested procedure vis-à-vis other therapeutic alternatives.

During the discussion, the patient learns, among other things, of the uncertainties involved in this decision and the limits of the treatment. It gives her some idea of the reality of her expectations and of the complexity of the medical decisions. She also gets the opportunity to express her wishes, worries, and preferences. It gives the doctor an idea of what is important for this individual patient and what are her subjective needs, which then can be taken in account when reconsidering his recommendation according to these specific, formerly unknown data. It also reveals any misunderstanding and misbeliefs that might affect the patient's decision, in a way that enables the doctor to correct them, and to deal, as far as possible, with the patients' irrational worries.

The Consequences of Informed Consent

From a legal point of view, autonomous authorization serves two main goals:

1. It authorizes the physician to touch the patient's body and give her treatment, which otherwise would be regarded as battery or trespass.
2. It enables the patient practice her autonomy as to the decision about what should be done to her own body. However, when the patients' informed consent is

obtained through a process of shared decision-making, as described above, it has several extralegal consequences:

(a) True communication occurs, by which the patient feels that she is being respected as a person, and the doctor gains her trust.
(b) Patient's expectations become more compatible with reality.
(c) Doctor and patient get to understand each other's consideration needs and motives, in a sense that enables them to reach an agreement as to the most preferable treatment for this individual patient in the circumstances.

These may be translated to:

1. Better chances to choose the most suitable treatment option for the patient at a given situation, taking in account all relevant bio-psycho-social elements, rather than biomedical ones only.
2. Better adherence by the patient to the treatment and her being more cooperative.
3. More realistic expectations are followed by less disappointment if anything goes wrong.
4. The patient shares responsibility to the outcomes, of which she was aware while choosing to undergo the procedure.
5. The patient cannot blame the doctor for an arbitrary, negligent decision.

Regarding the medicolegal consideration that a proper process of informed consent reduces the chances of the doctor to be sued, it would be worth citing things which were published at a symposium on professional liability of obstetrician-gynecologists [6]:

Informed consent is essentially a communication process and one that can prevent litigation—or at least successful litigation—when it is accomplished in a manner respective of the patient's autonomy. Regardless of how much time it takes, it is one of the best investments of time a physician can make toward avoiding eventual misunderstandings or mistakes that lead to litigation. Documentation of the process is critical.

Ethical Justifications of the Rules of Informed Consent: Respect for the Patient Autonomy

The prevailing ethical view, which has been widely adopted by the law, is that the primary function of informed consent is protecting and enabling individual autonomous choice [7].

As explained by the President's Commission, the foundation of this requirement is the fundamental recognition that adults are entitled to accept or reject healthcare interventions on the basis of their own personal values and in furtherance of their own personal goals [8]. In other words, we have a moral obligation to respect each other's autonomy, and doing things to other autonomous agents without their consent generally means overriding their autonomy. It is respect for people autonomy, or self-determination, that morally underpins the requirement of consent [9].

Indeed, the idea of self-determination in medical decision-making is not new. It had already been recognized on 1914, in the famous opinion of Judge Benjamin Cardozo who wrote the following: every human being of adult years and sound mind has a right to determine what shall be done with his own body; and a surgeon who performs an operation without his patient consent commits an assault for which he is liable in damages [10].

Informed consent introduces a new element to medical treatment. It is no longer a simple matter of consent to a technical assault; consent must now be based on a knowledge of the nature, consequences, and alternatives associated with the proposed therapy [11]. Providing the patient with all such information in a language that she can understand is supposed to give her the opportunity to make an intelligent choice, based on her own values and individual considerations and preferences. It also gives her some measure of control over her life as a patient, over what to be done to her body. By doing so, the doctor shows respect to the patient as an autonomous agent.

A Tool to Determine the Patients' Best Interests

Respect for the patient's autonomy is not the only consequence of informed consent. The other side of the coin is that the process of discussing the information with the patient to get her informed consent gives the doctor a chance to find out what are the patients' needs and values. In other words, it offers the possibility to determine the nature of this patient's well-being and what it actually entails, according to her values rather than the doctors or those of society at large. In this sense, the justification of the requirement of informed consent is the ethical principle of beneficence: it allows finding the ultimate option of treating the patient in a way that would serve that very patient's best interests.

It may be of interest to point out the position of the Jewish law as to patient's consent. As a religious law, it does not recognize the person's ultimate right over his or her body; moreover, it is a religious obligation to heal and to seek healing.

Therefore, the Jewish law does not recognize the right to exercise autonomy to the same extent as other ethical and legal systems [12]. Still, it gives much weight to the patient's guts feelings and her intuition regarding her body and self [13], while making therapeutic decisions in conditions of uncertainty, rationality, and medical knowledge alone cannot direct the decision-makers to the ultimate treatment. The choice must be assisted by intuition, both the doctors, which consists the art of medicine and the patients intuitions and feelings about the proposed treatment, even though, by definition, intuition is irrational (which does not mean that it is irrational to follow it) [14]. Hence, we can see that there is much more in the process of informed consent than mere respect to the patient's autonomy; it is necessary to determine the patient's best interests.

Practical Reasons for the Patient to Get the Information About the Treatment

After all, one point should always be remembered: the patient is the one who will have to live with the outcomes and consequences of the treatment. Therefore, she should be a partner to the decision-making and be given an opportunity to improve her chances to the best possible outcomes in the circumstances. Information is a key to this goal. Or in the words, it must be remembered that except in the rarest of circumstances, the ultimate decision belongs to the patient who has to live with the decision. One can only try to help a patient to make his or her best choice [15]. Having all the relevant information regarding the treatment is necessary in order to let the patient act in one of the following options:

(a) Consent to the proposed treatment.
(b) Prefer another form of treatment.
(c) Get a second opinion.
(d) Renounce treatment whatsoever.

Implications for the Fertility Patient

Though there are many kinds of fertility treatments, and rarely a one-shot solution, each trial is precious and should be considered very carefully. The patients' means - both physical, emotional and economical - are limited, and each failed trial diminishes them significantly. Moreover, time might be a critical factor, especially for women in higher ages. Furthermore, within various healthcare systems, patients' entitlement to subsidized reproductive treatment—and IVF in particular—is not unlimited; it might be restricted to a small number of cycles, so the patient must make sure that she gets the most promising option. Besides, in many countries, fertility patients can choose the doctor and/or clinic in which they will be treated, either because the healthcare system allows it or because this service is private. Therefore, it's essential that the patients receive full and reliable information regarding the proposed treatment, as well as all other options, their chances, burdens, and consequences, in order to make an informed decision. They can either consent to the proposed treatment or treatment plan, or prefer a different one which they find more suitable for them, or choose to consult or move to another doctor.

Challenges Related to Informed Consent in the Context of Assisted Reproduction

Reproductive therapy is a very unique medical field in the area of obstetrics and gynecology. Like obstetrics, it aims to assist patients to have babies. However, while obstetrics deals mostly with pregnant women, reproductive therapy deals with infertile patients, both female and male, and its main target is to achieve pregnancy resulting in a healthy baby. Reproductive therapy deals also with fertile people, who chose to donate their gametes or to carry a baby for others who cannot otherwise have children.

Fertility therapy has certain characteristics which complicate the informed consent:

(a) It uses various ever-expanding assisted reproductive technology (ART), each with its benefits, risks, and side effects, some of which maybe unknown, due to their novelty. Which means that with each medical advance in ART, informed consent for the associated therapies becomes more difficult, and the discussions grow increasingly complex [16].
(b) Procreative therapy decisions may have a significant impact not only on the patient undergoing treatment and the future offspring but also on the patient's partner, whether the gametes used for the process are theirs or donated. This raises hard questions like: Who is the patient? Who should give consent? Is the consent genuinely based on the patient's free will?
(c) Despite the desperate need of infertile patients, reproductive therapy is more elective than emergent. Therefore, and since multiple treatment paths may be reasonable, autonomy through informed consent is all the more important with ART [17].
(d) In the realm of fertility treatments, patients may receive information not only from their doctor, but from many other external sources such as friends, peers, and social media. As indicated, it's a "highly motivated patient population calling upon" Dr. Google [18].
(e) Reproduction treatments are intended to create offspring in ways and circumstances, which occasionally raise severe ethical and legal consequences regarding the well-being of the future child, as well as establishing and relinquishing parental rights.
(f) Specifically, the in vitro fertilization (IVF) process is a multi-level process that involves several ART procedures: ovulation induction, gamete collection, in vitro fertilization, embryo freezing, and embryo transfer, each of which entails critical decision-making. Thus, informed consent to IVF is ongoing throughout the process, which contains a continuum of decision points. Moreover, it does not just relate to a one-off medical procedure—it is also about processes with future implications, such as the storage of gametes (sperm and eggs), the storage of embryos, and how gametes or embryos may be used in the future [19].
(g) Infertility and its therapeutic process entails unique emotional, psychological, and social implications, on top of the physiological ones [20]. These special characteristics imply that the informed consent process (hereinafter IC) to ART is unique in nature. It should cover lot of quite complex information, related to several procedures, some of which entail hard value decisions that must be made autonomously by the patients according to their values. This requires a separate discussion with the patients at each decision point, explaining all the

relevant details, regardless of the general IC at the initial consultation and the following signature on the informed consent form.

Patients usually have former knowledge about the treatment from various sources, which may be erroneous, so the doctor has to correct it and convince the patients to trust him. It might, on the other hand, be accurate information relating to advanced procedures or add-ons which may or not be relevant to these patients' case [21]. In such situations, the doctor, again, ought to explain and negotiate with the patients their suggestion (and be humble enough to accept it, when appropriate, though it was not his idea). Additionally, a special attention should be given to the effect of the treatment on the patient's emotional state.

Informed Consent to ART in Practice

Many fertility professionals and lawyers have expressed concern over whether patients understand and appreciate the implications of ART for personal health and the health of children born through the process. Some of the disturbing questions relate to the lengthy, jargon-filled informed consent documents that might be too complex or intimidating. Do patients read these papers before they sign them?

Are Patients Simply Overwhelmed with Information?

Another disturbing question relates to the patients' state of mind: Are they so eager to conceive a child that they discount the gravity of the risks or low probability for a desired outcome [22]?

These questions and a lot more were widely researched by the scholar Jody Lynee Madeira, a law professor at Indiana University. Her research found that the vast majority of patients read the forms and that at least in IVF recall and comprehension are not problematic. However, they prioritize physician conversations and may place higher value on information learned face-to-face. Patients like it best when physicians explain why certain information is crucially important for their personal well-being. Hence, the authors conclusion is that (their) findings support renewed efforts to make informed consent more personal and meaningful, and less bureaucratic, to patients. Doctors should explain to patients why forms are relevant, but realize that information delivered through interpersonal conversations is more impactful. Such patient engagement also allows treatment teams to most efficaciously assess patient understanding [23].

Furthermore, for the vast majority of patients, signing consent documents did not mark the beginning or ending of the IC. Conversation played a large role in patients IVF consent interactions; nearly all the surveyed patients found consent conversations more or equally helpful than forms [24]. The experience of lack of informal discussion of consent information had been described by a patient; if we had had a better relationship with our doctor, I would've been able to make a significantly more informed consent. His attitude was like, 'It's all there on that paper.' . . . I felt incredibly informed about the process in general, but not to my specific case [25]. The search for answers about their specific situation is a common reaction of patients after the treatment started, even though at the initial consultation they had received detailed information about the treatment, its possible risks and so on. They want to know: how does the information fit my/our situation? [26].

It should be noted that embryo disposition forms are a different issue; the decision to undergo IVF and embryo disposition choice are two different tasks; the former is likely already made at the time of consent, while the latter forces patients to consider possibilities that are novel, personal, difficult, and even quasi-parental [24].

For the patients, signing embryo disposition forms is a very different experience than completing IVF consent documents. Many surveyed patients had described the need to make decisions about frozen embryos that did not yet exist, and maybe never would, as surreal and upsetting, and several felt it was too early to choose dispositions [24]. Even from a legal perspective, it is not clear whether forms containing patient dispositional decision-making should be viewed as legal agreements or medical consent forms. The American Society of Reproductive Medicine (ASRM) clearly recommends separation of the medical informed consent for IVF treatment from a stand-alone contract for disposition of embryos, including discard as a default disposition provision if other selected options are not available [27].

In recent years, multimedia interventions were added to the IC. Studies linked these interventions to greater patient enjoyment; improved patient knowledge, comprehension, and recall; improved physician-patient relationships; and lower anxiety and faster learning. However, these aids should not be viewed as substitute for interpersonal interaction with patients; rather, they can help both parties prioritize issues, promote dialogue over monologue, and assess and refine understanding [28].

Litigation regarding informed consent in reproductive therapy remarkably, as far as can be concluded from review of the case law in the USA and in Israel, there are barely few legal cases against fertility doctors, related to informed consent. According to comprehensive review of the American case law relating to ART [27], the issue of consent was raised mainly in litigation concerning the use of frozen embryos and reproductive tissue, in circumstances of divorce or after the death of one partner. Yet, a legal issue concerning informed consent, that has been raised recently, relates to pre-implantation genetic testing ("PGT") of IVF embryos, more specifically how to counsel and obtain informed consent from patients who may only have "mosaic" embryos (a more nuanced emerging categorization between normal and abnormal) given as yet unanswered questions as to both how representative a single cell may be and—after some children from mosaic embryos were born without abnormalities—whether a self-correcting mechanism may be at play [27].

The issue of PGT had been raised also in an Israeli case [29]. The patients conceived via IVF at 2006, and the child turned out to have Down syndrome. They claimed that the doctor should have informed them, before embryo transfer, about the option of the procedure known at the time as pre-implantation genetic selection ("PGS"). This especially since their religion (Islam) does not allow amniocentesis nor abortion. The suit was denied, since the patient had no known risk factor, while at that time PGS involved immediate and probably long-term risks to the embryo and had been done only in very rare indicative cases.

Three other Israeli cases are related to informing and consulting the patients about the number of embryos to be transferred. In one case [30], the court ruled that the defendant doctor had violated the plaintiffs autonomy in three ways:

(a) By not telling them about the results of the fathers' tests, according to which he had varicocele nor about the treatment options of this medical problem (the defendant explained this omission claiming that the father had a low sperm count, so operation would not have been effective)
(b) By avoiding to get the fathers written informed consent for the embryo transference
(c) By his omission to inform the patients about the correct number of embryos transferred to the mother womb (7 rather than 6, as reported to the patients)

Three embryos had been implanted, and following reduction of one, the pregnancy resulted in the premature birth of twins, one of whom severely handicapped. However, the court determined that the patients would not have done anything different, since they trusted the defendant and followed his advice with no hesitance. As they admitted, their state of mind was that "all you want is to achieve pregnancy", so they were ready to take the risk of multi-fetal pregnancy resulting in premature delivery with its entire consequences. Therefore, the plaintiffs were granted compensation just for the violation of their autonomy, but not for negligence (since there was no causal connection between the doctor's omission to inform the patients and the tragic result).

In the second case [31], the patient underwent IVF treatment, during which three embryos were transferred to her womb. At the sixth week of pregnancy, the plaintiffs advised refused reduction from religious and conscious reasons. The triplet had been born prematurely; on the 31 weeks of pregnancy, one of them severely damaged. The court accepted the parents and claim that they had not received any explanation at any stage before the embryos transfer. They were not informed about the risks involved in a triple pregnancy nor about the need of embryo reduction if all three embryos were implanted. Neither were they consulted regarding the question how many embryos to thaw. The court described the case as one of a young innocent woman, with no life experience and no knowledge about the process of IVF, who had trusted the doctors, had not asked questions, and is contented with the information they had given her, having no idea that it had been lacking partial and somewhat inaccurate, excluding mandatory information about risks that should have been explained to her and her husband.

In the third case [32] the plaintiffs had undergone IVF treatment at 2000, during which five embryos had been transferred to the mother's womb. The patients refused multi-fetal pregnancy reduction, and the pregnancy ended on the 23 weeks, in the birth and death of three very small premature babies. The plaintiffs claimed that the doctor had not informed them about the consequences and risks involved in transference of five embryos. The suit was denied by the court that believed the doctor who proved that he had explained and discussed with the patients the risks of multi-fetal pregnancy, including abortion, death, and pre-maturity, and therefore the need of reduction. The court concluded that the plaintiffs knew that transference of five embryos might end up in multi-fetal pregnancy which can result in a premature birth, and they chose it while being aware to the risks, due to their intense will to become parents. Their informed decision to consent to the transference of five embryos had been made in order to increase their chances for pregnancy.

Lessons to Be Taken from the Case Law

As we can see from the facts of the above cases, the transference procedure entails critical decisions: which and how many embryos should be transferred? These decisions have impact on the chances to achieve pregnancy and to have a baby or babies, but much more so on the entire life of the future child or children to be born. Thus, these are not pure professional decisions, but rather value judgments related to the quality of life of the patients and the future children. Therefore, these decisions must be considered very carefully together with both the patient and her partner—the parents to be. Furthermore, the doctor should not take for granted the option to correct the situation of multi-fetal pregnancy by reduction. Reduction is a moral issue, like abortion, that has serious ethical and psychological implications, as well as religious ones. It is much more complicated for ART patients, who were striving so hard to have the fetus(es) which they are now required to abort. Therefore, the IC must include thorough explanations about the decisions that should be made prior to embryo transference. Since the transference decisions might seem remote for patients who have just consented to undergo IVF, it would be necessary to explain and discuss it with them again after egg retrieval and upon a successful fertilization of eggs, before the transference process. Remarkably, the phenomenon of patients' choice to transfer more embryos in order to increase their chances to conceive, regardless of the risks of multi-fetal pregnancy, might be more common than expected. An Israeli study examined empirically the impact of the IC on patient knowledge and understanding of potential hazards associated with IVF treatment [33]. Interestingly, the authors' hypothesis that patients' better understanding of potential complications would be translated and expressed as rational choices of treatment alternatives had been disproved. Just 25.5% [12] of the study group (women after the IC) considered delivery of a single baby as their optimal result, compared to 32.6% [15] of the control group (before IC). Furthermore, preferences shifted toward triplets: eight patients (17%) after IC considered this option as their best result, compared to only five patients (11%) before IC. Paradoxically again, after IC, no woman in the study group wanted to have a single embryo transfer, compared to 5 (11.9%) in the control group (before IC). Furthermore, 26 women (56.5%) in the study group wanted to transfer 3 or more embryos compared to 14 (33.3%) in the control group.

Still, it should be noted that the IC takes place before initiation of the treatment, and it covers the entire process. So, the study group does not necessarily represent patients at the transference decision point. However, this study reinforces the need to go through the relevant information regarding embryo transference when it is actual.

In general, the extremely small number of lawsuits related to informed consent to reproductive treatments, both in the U.S. - the "leader" of medical malpractice litigation—and in Israel—the country with the highest percentage of IVF procedures per population in the world [34]—calls for the conclusion that the need to improve and deepen our understanding of this topic is not grounded in medicolegal concerns. Rather, we should examine it from the patient's perspective.

Informed Consent from ART Perspective

As explained at the beginning of this chapter, IC is a process of shared decision-making. If done properly, it should include, on top of the medical information delivery, an opportunity for the patient to express her wishes, worries, and preferences so the doctor can get an idea of what is important for this individual patient and what are her subjective needs, which then can be taken in account when reconsidering his recommendation. So, IC is the heart of patient-centered care (PCC), which is defined as the provision of treatment that respects and responds to the patient's personal values, preferences, and needs and ensures that clinical decisions are guided by the patient's values [35].

Patients undergoing fertility care are particularly likely to benefit from the PCC approach, due to the heavy physical and psychological burden imposed by fertility problems and treatment and given their deep emotional implications. PCC has been found as central in how couples experience infertility treatment. A Dutch study even reports patients' willingness to trade off a higher pregnancy rate in order to receive more patient-centered care, and it mentions the lack of patient-centeredness as the most cited non-medical reason for changing fertility clinics [36]. Interestingly, the value patients attached to the patient-centeredness of care is remarkable and significantly higher than physicians would recommend: patients were willing to trade-off up to a third (9.8%) of pregnancy rate for more patient-centered care, whereas physicians recommended to trade-off up to 6.3% [36].

Research of the implication of the PCC approach in fertility treatment in Israel revealed some very interesting facts about the gaps between patients and directors and staff comprehension of the patients' needs and their fulfillment [35]. Interviewed unit directors familiar with the PCC approach implement it at some level in their work and in most cases support it. Most also recognize that in order to provide PCC, it is important that the patients have access to the staff, that they be given information and explanations, and that they be involved in the treatment. However, they placed less emphasis on other dimensions of PCC, such as respect of the patient's values and needs and the communication skills of the staff. The interviews also revealed that only some of the units had a social worker or psychologist on their permanent staff.

The patients gave the highest scores to the dimensions of staff communication skills (2.3 (out of 3)) and the professional competence of the staff (2.3). In contrast, emotional support received the lowest score (1.0). Information and explanations, respect of the patient's values and needs, and accessibility to staff received relatively low scores as well. Differences were found between the scores the patients gave and the scores given by the staff members. In terms of information and explanations, respect of values and needs, and emotional support, the patients gave a score for the unit in which they were treated that was lower than that given by the staff members working in the same unit. The gap in the dimension of emotional support was particularly large (in this dimension, the staff gave a score of 2.5, while the patients gave a score of 1.0).

The dimension of emotional support included four items; one is opportunity to consult mental health professional (psychologist or social worker) familiar with the area of reproductive therapy, who gives emotional support; and the three other items relate to providing information:

(a) Information about potential emotional impact of the treatments
(b) Information about support groups for women undergoing fertility treatments
(c) Information leaflet about the fertility treatments for the partner or relatives.

In all of these items, there were huge gaps between patients and staffs scores. About 80–90% of the staff participants gave these items the highest score (3) vis-à-vis 26–38% of the patients. Furthermore, the unit directors had not mentioned the issue of information to the patients about possible emotional impacts of the treatments. (Neither had they mentioned discussion with the patients of the prospects of treatment success and the willingness to discuss test results or treatment errors.)

The patients felt that in relation to everything connected to emotional support from the staff, there is not enough focus on their needs. They were not asked if and to what extent they would like to receive various components of the treatment, among which emotional support. It is also worth mentioning that at the units which had social worker as part of the staff, the emotional support scores were particularly high.

Emotions affect everyone undergoing infertility, but they differ from person to person and often unpredictable. Emotions, for better or worse, play key decision-making roles. In providers' experience, emotions affect patients in complex and contradictory ways, steering them toward or away from treatments. Intense feelings might motivate some patients to pursue more aggressive treatments early on, or to be aggressive with embryo transfers, or attempt to cycle too soon after an unsuccessful attempt, or attempt any treatment option. Yet, once they have thought it through, most patients become more rational. Conversely, emotions also deter patients from using certain medications or moving on to more advanced treatments and may even drive them to prematurely cease treatment [37]. I meant to say that according to the above findings, it is obvious that the patients need that their emotional difficulties should be addressed by the professionals as an inherent part of the IC.

It should be noted that on a webinar held by Cooper Surgical on fertility informed consent in the digital era, the participants, Dr. Steven Lindheim, Dr. Jody Madeira, and Dr. Linnea R. Goodman, agreed that given the level of anxiety, awareness on psychological resources is very important. Patients should know that it is ok to have mental health help and support should always be offered [38].

Conclusion

Reproductive medicine is unique; it operates at the intersection of sick and healthy. The majority of infertility patients do not perceive themselves as "sick." According to some sources, they are healthy adults in need of some assistance to become pregnant [26]. Yet, others claim that women experience infertility as a chronic medical condition and have psychological symptoms equivalent to those of patients with cancer, cardiac issues, and hypertension [36]. It is an ongoing multi-level process, consisting of various procedures which involve decisions that demand high professional specialty, but also crucial subjective value judgment of the patients. It deals with critical problems elective. The patients' medical problems are physical, but their implications are psychosocial.

All these aspects should be reflected and considered in the informed consent process. The doctor should be led by the understanding that signing the informed consent form following the initial consultation is just the beginning of the IC. Informed consent is ongoing throughout the in vitro fertilization (IVF) process, which contains a continuum of decision points [19]. Consequently, the doctor should explain and discuss with the patients the relevant information related to every procedure before its execution and share the decision-making with them, honoring their values. Furthermore, all throughout the treatment process, he should give the patients information regarding their specific case. This is extremely important for patients, especially after they comprehended the general information related to the diagnosis and treatment.

The name of the game is patient-centered care, and the heart of which is the informed consent process.

References

1. Salgo v. Leland Stanford Jr. University Board of Trustees, 317 p. 2d 170 (Cal. Dist. Ct. App. 1957).
2. Kirby MD. Informed consent: what does it mean? J Med Ethics. 1983;9:69–75.
3. Gillon R. Philosophical medical ethics, 113.
4. Beauchamp TL, Childress JF. Principles of biomedical ethics, 3rd ed. 1989. p. 76.
5. President Commission for the study of ethical problems in medicine and biomedical and behavioral research, making health care decisions. Government Printing Office, Washington, D.C. 1982. p. 2.
6. Professional liability. Clinical Obstetrics and Gynecology, 1988, vol. 31, no. 1, 139–168, Daniel K. Roberts, "Prevention: Patient Communication", at p. 158.
7. Beauchamp TL, Childress JF, supra note 4, at 75.
8. President's Commission, supra note 5, at 3.
9. Gillon, supra note 3, at 114.
10. Schloendorff v. New York Hospital, 211 N.Y. 127, 129, 105 N.E. 92, 93 (1914).
11. Mason and McCall Smith, law and medical ethics, 4th ed., Butterworths, 1994. p. 238. See also Brazier M, Patient autonomy and consent to treatment [1987] 7 Legal Studies, 169.
12. Steinberg A, Jewish A. Perspective on the four principles. In: Gillon R, editor. Principles of health care ethics. 1994. p. 68–70; Steinberg A. Encyclopedia of Jewish medical ethics, Vol. II, Informed consent, p. 22–33.
13. Following the verse in Proverbs, 14, 10 The heart knoweth its own bitterness.
14. Patients' intuitions will be discussed further in this chapter. For the part of intuition in decision making by doctors see Atul Gawande, Complications, Metropolitan Books 2002, 228–252, and the sources to that chapter.

15. Katz J. The silent world of doctor and patient, (1984). The Johns Hopkins University Press, 2002. p. 128.
16. Madeira J, et al. Using the engaged MD multimedia platform to improve informed consent for ovulation induction, intrauterine insemination, and in vitro fertilization. Fertil Steril. 2018;110(7):1338–46.
17. Madeira JL, et al. Inform and consent: more than just. Fertil Steril. 2017;108(1):40–1.
18. Fertility informed consent in the digital era. Cooper Surgical Journal Club. 2020. https://fertility.coopersurgical.com/webinars/fertility-informed-consent-in-the-digital-era/.
19. ACART Advisory Committee on Assisted Reproductive Technology. Informed consent and assisted reproductive technology, Proposed advice to the Minister of Health. Consultation Document, 2015.
20. Madeira JL. Taking baby steps: how patients and fertility clinics collaborate in conception. University of California Press; 2018. https://www.jstor.org/stable/10.1525/j.ctt1ws7vt8.
21. Iacoponi O, van de Wiel L, Wilkinson J, Harper JC. Passion, pressure and pragmatism: how fertility clinic medical directors view IVF add-ons. Reprod Biomed Online. 2022;45(1):169–79.
22. Madeira JL, et al. Inform and consent: more than just "sign". Fertil Steril. 2017;108(1):40–1.
23. Madeira JL, Christianson MS, Lee M, Preston Parry J, Pennings G, Lindheim SR. Inform and consent: more than just "sign here". J Reprod Med. 2020;65(4):213–9.
24. Madeira JL. The ART of informed consent: assessing patient perceptions, behaviors, and lived experience of IVF and embryo disposition informed consent processes (2015). Articles by Maurer Faculty. 1953. https://www.repository.law.indiana.edu/facpub/1953.
25. Madeira JL. Taking baby steps: how patients and fertility clinics collaborate in conception. University of California Press. 2018. chrome-extension://efaidnbmnnnibpcajpcglclefindmkaj/http://petrieflom.law.harvard.edu/assets/publications/Madeira_Informed_Consent_Reproductive_Medicine.pdf.
26. Kvernfaten B, Fedorcsák P, Solbrække KN. It's all about kids, kids, kids! Negotiating reproductive citizenship and patient-centred care in 'factory IVF'. BioSocieties. 2022.
27. Crockin SL, Altman AB, Edmonds MA. The history and future trends of art medicine and law. Fam Court Rev. 2021;59(1):22–45.
28. Madeira JL, et al. Inform and consent: more than just "sign here". Fertil Steril. 2017;108(1):40–1.
29. Civil case 19055-12-13, Jerusalem District Court (10/2/2016), published on https://www.psakdin.co.il (in Hebrew)
30. Civil case 1499-02, Tel Aviv District Court (17/12/2008) published on https://www.psakdin.co.il (in Hebrew), and Civil Appeal 980/09, The Supreme Court (15.3.2012), published on ruling.co.il (in Hebrew)
31. Civil case 57625-01-13, Tel-Aviv District Court (1/8/2017), published on https://www.psakdin.co.il (in Hebrew)
32. Civil case 3827-12-07, Central District Court (24/08/2011), published on https://www.psakdin.co.il (in Hebrew)
33. Voskoboynikov-Ugortsev I, Rosenfeld Y, Keinan Boker L. In vitro fertilization informed consent: revisited, empirically. Isr Med Assoc J. 2020;11(22):681–3.
34. IVF and fertility treatment at Rambam hospital https://www.ihospitalonline.org.il/ivf-in-israel.aspx
35. Artom T, Hass B-S, Zhivaev L, Implementation of the patient-centered care approach in fertility treatment in Israel. Myers-JDC-Brookdale Institute, Jerusalem 2019. https://brookdale.jdc.org.il/en/publication/patientcentered-care-fertility-treatment/
36. Van Empel IWH, Dancet EAF, Koolman XHE, Nelen WLDM, Stolk EA, Sermeus W, D'Hooghe TM, Kremer JAM. Physicians underestimate the importance of patient-centredness to patients: a discrete choice experiment in fertility care. Hum Reprod. 2011;26:584–93.
37. Madeira JL. Taking baby steps: how patients and fertility clinics collaborate in conception. Berkeley: University of California Press; 2018. CHAPTER 1 Consent and sensibility: emotions, decision making, and informed consent in reproductive medicine. p. 21–48.
38. Informed Consent in the Digital Era Cooper Surgical Journal club 07/08/20. https://fertility.coopersurgical.com/webinars/fertilityinformed-consent-in-the-digital-era/

Part IV

Innovative Technologies

Artificial Intelligence in Reproductive Medicine

14

Assaf Ben-Meir and Natali Schachter-Safrai

Introduction

The main goal of reproductive medicine is delivery of a healthy newborn. This journey includes many stages: evaluating couples experiencing infertility, calculating the chances for successful treatment, deciding on the optimal treatment protocol, monitoring and fine-tuning ovarian stimulation through treatment phases until oocyte retrieval, evaluating the potential of the retrieved oocytes and sperm, assessing embryo quality and predicting implantation potential, evaluating endometrial receptivity, and determining the correct time for embryo transfer (Fig. 14.1). All these stages can be assisted by artificial intelligence (AI), a domain of computer science which is rapidly evolving in many areas, including the reproductive field. AI offers certain advantages which might increase IVF efficiency, including consistent decision-making based on data and the ability to integrate and analyze multiple and complex variables. In addition, AI can be incorporated into the IVF laboratory workflow and improve quality control. In this chapter, we will review some of these aspects in which AI has begun to gain traction in reproductive medicine.

A. Ben-Meir (✉) · N. Schachter-Safrai
IVF Unit, Department of Obstetrics and Gynecology, Hadassah Medical Organization and Faculty of Medicine, Hebrew University of Jerusalem, Jerusalem, Israel

J. G. Schenker et al. (eds.), *Hot Topics in Human Reproduction*, Reproductive Medicine for Clinicians 3, https://doi.org/10.1007/978-3-031-24903-7_14

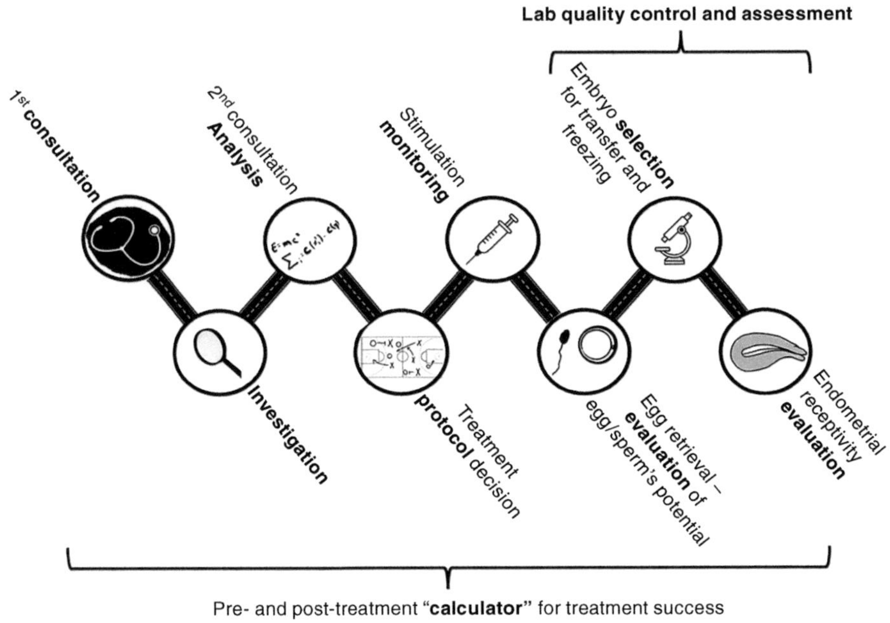

Fig. 14.1 The IVF journey. Artificial intelligence can assist in each step and analyze the whole process for quality control and prediction of success rate

How Does It Work?

Artificial intelligence (AI) is a branch of computer science which involves developing technologies capable of imitating human intelligence.

Machine learning is a form of AI in which a computer analyzes a vast amount of data, builds algorithms which find patterns based on the data, and implements the algorithms in different situations [1]. The process is developed using three datasets: training, validation, and test.

One of the methods used in the training process is "deep learning," in which multiple layers of artificial neural networks, simulating human neurons, are used. This method helps in refining raw data processing. The learning process in the training dataset may be carried out either by supervised (with labeling of the data such as implanted or not implanted embryos) or unsupervised learning. Thereafter validation takes place by testing the algorithms on a separate subset of the data (called the "validation set") to measure the output error and choose the most optimized algorithm. Finally, the final algorithm is tested over a third and separate dataset called a test dataset, to verify performance.

AI Limitations

Being a derivative of the data presented, the performance of an AI model is associated with the quality of the database on which it was trained. AI models are usually based on Big Data, since the size and quality of the database used to train an AI model is essential for its success. The larger, wider, and more diverse the spectrum of data on which the model is based, the more generalized and clinically applicable the algorithm will be. Furthermore, training an algorithm on an "unbalanced" data (meaning there is bias in numbers of positive and negative examples) might create a bias by its tendency to recognize parameters that are in the majority. Another possible limitation, referred as "overfitting," relates to a model which narrowly specializes in the data on which it was trained and thus loses its abilities to generalize an algorithm to new data.

The implementation of the above principles is especially challenging in the reproductive field due to the restricted numbers of available embryos for evaluation. Therefore, multicenter collaborations have been established in order to overcome these challenges by enlarging and diversifying the AIs' databases.

Implementation of AI in IVF

AI and Monitoring During Treatment

The management of ovarian stimulation during an IVF cycle is challenging. Decision-making regarding the duration of stimulation, gonadotropin dosage alterations during the cycle, and the timing of ovulation trigger or cycle cancellation are key points for consideration during the IVF cycle. The use of AI in this field is scarce due to complexity and subjective nature of such decisions.

In their work, Letteri et al. have developed a computer algorithm designed for IVF management and assessed its accuracy in decision-making during IVF treatments when compared with evidence-based decisions by the clinical team [2]. They demonstrated high accuracy and concordance of the proposed algorithm decision with clinical decisions regarding four key decisions (stop/continue stimulation, trigger/cancel, days to follow up, and dose adjustment) during the process of ovarian stimulation. They concluded that "These tools are no substitutes for hands-on patient care but can be added to the decision process to optimize outcome."

Imaging Analysis of the Ovary and Uterus

One of the most successful uses of AI in medicine is in imaging analysis. The clinical utilization of AI in medical imaging is used for image segmentation (recognition and segmentation of the regions of interest), feature extraction (such as morphological and texture features), and definition of classification systems [3].

Transvaginal ultrasound (TVUS) is an essential diagnostic tool for women undergoing fertility treatments, as it helps to assess the ovarian reserve and follicle development monitoring. The measurement of the antral follicle count (AFC) by ultrasound is used for the appraisal of the ovarian reserve and is useful for tailoring the fitting stimulation protocols. AFC is prone to inter- and intra-observer variability, consequently ideal for AI implementation. Follicles can be identified either by using various segmentation techniques based on the areas of detected follicles (such as pixel intensity level) and features of the image (such as roundness) [4] or by performing echotexture analysis [5].

Once trained to assess ovarian follicles, the AI system can also monitor the ovarian response to ovarian stimulation. Robertson et al. demonstrated that AI could help to determine the frequency of follicle tracking during stimulation in order to predict trigger day and risk of hyper-response. The integration of ultrasound measurements of the ovarian response together with other variables, such as estradiol levels, gonadotrophin dosage, and day of stimulation, may further help in managing the ovarian stimulation cycle and GT adjustments throughout its course [2].

Another possible implementation of AI in infertility imaging is endometrial assessment. Endometrial thickness is measured as part as one of the biomarkers for endometrial receptivity and the chance for embryo implantation. Besides endometrial thickness, which can be evaluated more rapidly and objectively, ultrasound image processing can utilize additive information including endometrial regularity, volume, appearance (trilaminar or hyperechoic), and sub-endometrial vascularity and motility [6]. Integration of all these parameters can improve our understanding and prediction of endometrial receptivity before embryo transfer.

Sperm and Oocyte Potential Analysis

The use of AI prediction in the evaluation of oocyte competency is of significance when facing clinical decisions regarding oocyte selection for fertilization (especially in countries where legislation restricts embryo selection), blastomere or trophectoderm biopsy for PGT, and embryo transfer. The assessment of the oocyte in the setting of ART is traditionally performed prior to IVF by the morphologic evaluation of the oocyte–cumulus complex that may provide data on oocyte maturity. No previous studies have evaluated the use of AI for the assessment of the oocyte–cumulus complex for prediction of oocyte maturity. However, several studies focused on analysis of oocytes that underwent removal of their cumulus prior to ICSI. A study by Cavalera et al. used AI for the analysis of mouse oocyte images collected from time-lapse imaging systems while performing in vitro maturation of GV oocytes to M2 oocytes [7]. This study observed the profile of cytoplasmic movement velocities throughout the culture period and used mathematical classification tool (feed-forward artificial neural network, FANN) to predict the probability of a gamete to be developmentally competent with high accuracy.

The ability to predict oocytes quality using AI was primarily demonstrated in a study by Manna et al. [3]. This study investigated the use of neural network for texture analysis of oocyte images from two datasets—of oocytes and embryos—and showed morphology classification performance for oocyte quality. Later studies focused on specific oocyte morphologic parameters such as oocyte cytoplasm and polar body characteristics as well as features of the zona pellucida and the cumulus cells for semantic oocyte segmentation [8]. They were able to predict oocyte quality by testing the association of oocyte morphology with pronuclear development and the subsequent embryo development.

Semen parameters analysis has also been a target for the use of AI to improve efficiency and accuracy of sperm selection in various male infertility diagnoses. Although the computer-assisted semen analyzer (CASA) methods enable partial automation of semen analysis using low-level machine learning, their challenging application acts as a major drawback for wider use. In a study by Javadi and Mirroshandel [9], 1540 sperm images from 256 infertile men were analyzed for detection of morphological deformities in acrosome, head, and vacuole abnormality using a deep learning detection method. Their proposed algorithm achieved $F_{0.5}$ of 84–94%. The potential for detection of sperm DNA integrity using AI was also demonstrated [10]. Deep convolutional neural network ability to predict DNA quality from brightfield images was shown in a system trained on a collection of 1000 sperm cells of known DNA quality. This system demonstrated moderate correlation between a sperm cell image and DNA quality and the ability to identify higher DNA integrity cells, similar to the current manual microscopy-based sperm selection.

In a study by Mirsky et al. a model for classification of sperm morphology ("good" or "bad") was developed through use of interferometric phase microscopy with support vector machines reaching a high 88% accuracy [11]. In a study by Goodson et al. an automated method that classifies patterns of sperm motility during in vitro capacitation following the removal of seminal plasma was developed [12]. Sperm motility was assessed by CASA using a support vector machine-based decision tree to compute and separate five sperm motility classes with a nearly 90% accuracy. This system classification provided a quantitative method for monitoring alterations in sperm motility.

Interestingly, AI application for sperm analysis has been additionally explored in the setting of smartphone-based applications used to determine the functionality and maturity of sperm, as well as sperm viability and DNA integrity in fresh semen samples [13, 14].

In severe male infertility cases, one of the major challenges is acquisition of sperm in males with azoospermia. In such cases, surgical testicular extraction of sperm is performed, and the attempt to identify the different types of spermatozoa by the embryologists and differentiate them from other tissue cells is incredibly challenging and time-consuming. The ability to accurately identify sperm cells for ICSI in these cases is on the main challenges in the field of AI and severe male infertility, with paramount beneficial potential for clinicians and embryologists.

Embryo Selection

Embryo assessment is traditionally done manually by analyzing embryos under a microscope and assessing different parameters concerning their morphology. This subjective technique is prone to inter- and intra-operator variability [15]. Moreover, the implantation prediction of morphology alone during the cleavage stage is low [16]. The emergence of time-lapse incubator (TLI), which enables detailed morphologic and kinetic evaluation using automated image capturing every 10–20 min, has provided new possibilities in embryo assessment. Morphokinetic assessment of the embryo includes annotation of the time of each cell cleavage and embryo milestone development. Several studies tried to incorporate this data into algorithm to predict embryo implantation with limited success [17]. Moreover, the manual annotation significantly increased the workload on embryologists. AI-based automatic image processing and annotation can be performed, allowing to standardize this process and assist in embryologist workflow. In addition to annotating the time of each cell cycle, the system must be capable to identify other features essential in embryo evaluation, including morphology (fragmentation, blastomere symmetry, stage of expansion for blastocyst, and evaluation of inner cell mass and trophectoderm) and cell-division abnormality (direct uneven cleavage, reverse cleavage). Integration of all this information will improve and accelerate machine-learning capability.

Deep learning methods, and more specifically convolutional neural network (CNN), are the methods by which visual information is processed. CNN can be used for automatic cell annotation with high accuracy [18, 19], embryo grading and selection [20], and blastocyst and implantation prediction [21].

Assuming that IVF outcomes are the result of complex interactions between known and yet unknown parameters, others chose to use deep learning models to analyze the entire raw time-lapse video without using annotated parameters, making use of every data point collected from time lapse to predict the probability of fetal heart pregnancy [22].

Currently, the implementation of an AI model in different IVF labs is restricted by the diversity between the labs, including different TLM or medium culture. Therefore, the models should be assessed for generalizability and be interpreted with caution.

Ploidy Prediction

The gold standard for embryo selection based on ploidy is achieved by preimplantation genetic testing for aneuploidies (PGT-A). However, this technique involves invasive removal of cells and requires special equipment and embryologist training. Therefore, research efforts are focused on noninvasive techniques for selecting euploid embryos. Several studies have tested the association between morphokinetic parameters and embryo ploidy [23, 24], but sufficient accuracy for replacement of PGT-A was not achieved. The use of TLI-based AI models for prediction of ploidy was introduced by Chavez-Badiola et al. [25], in a study that tested AI models for embryos ranking to predict euploidy. Their model was able to sort blastocysts based on their predicted ploidy during embryo selection with a prediction accuracy of 0.70.

Recently, Huang et al. [26] used TLI data from euploid and aneuploid embryos and were able to develop an euploid prediction algorithm. This AI-based model predicted euploidy accurately, with an area under curve (AUC) of 0.80.

Quality Control in the IVF Lab

Quality control in the IVF lab is critical, as lab performance was implicated directly with treatment results [27–29]. The control must be measurable, and some key performance indicators (KPI) were formally defined in the Vienna consensus document [30]. The potential of such KPIs to improve IVF lab quality depends on objectivity of such measurements, accurate recording, and the time till measured results and analysis to get correction in reasonable time.

Application of AI in the IVF laboratory might decrease embryologists' variability, by reducing manual procedures and subjective assessments. Bormann and colleagues examined two automatically measured KPIs on IVF results: fertilization rate after intra-cytoplasmatic sperm injection (ICSI) and day 3 embryo quality [31]. In their study, they trained the computer on 2366 embryos to identify 2 pronuclear appearances (normal fertilization) and test it on 947 embryos with prediction accuracy of normal fertilization of 93.1%. They also revealed an algorithm for categorizing embryos into blastocyst and non-blastocyst with 90.2% accuracy. They found association between those measurements and pregnancy rates. Moreover, they showed that such system could detect in advance a drop of pregnancy rate below 50%.

AI may also control for different environmental aspects relevant to embryo development. Environmental sensors pave the way to standardization of quality control in the IVF lab (Maninder), not only by detecting changes in light, room temperature, and humidity but also by transmitting the information to computers in which the data is collected, stored, analyzed, and used to instruct other devices how to respond to environmental changes. Furthermore, the ability to monitor embryo development in relation to different environmental parameters (including culture media) might help to make the relevant changes for optimal outcomes.

AI and Pregnancy Prediction/Success IVF

The final goal of IVF treatment is live birth of a healthy newborn. The journey to a successful cycle incorporates emotional, physical, and financial burden. As a result, prediction of the chances for successful treatment in the next cycle is crucial for patients, clinicians, and policy makers. Although women's age is maybe the most single predictor for pregnancy and live birth, it is not the only one, and a more personalized prediction is mandatory.

Previous studies tried to produce such pregnancy prediction calculator. McLernon and colleagues used the UK national data from the Human Fertilisation and Embryology Authority (HFEA) register between 1999 and 2008 [32]. The dataset included 113,873 women with 184,269 complete cycles. They used a discrete time logistic regression model to predict the chance of a live birth after a maximum of six cumulative complete cycles of IVF or ICSI. Key predictors of live birth were women's age, duration of infertility, number of eggs collected, cryopreservation of embryos, and stage of embryo transferred. The concordance index of the model was around 0.72–0.73. The models have been converted into an online calculator (https://w3.abdn.ac.uk/clsm/opis). The main advantage of the study was the huge number of cycles in the dataset. The disadvantages included limited availability of other potentially important predictors not included in HFEA database, no comparison to other machine learning or deep learning algorithms, and manual backward selection process of incorporating "must" predictors. Nevertheless, a systematic review of the quality of clinical prediction models in IVF found this calculator with the best performance [33].

In the last 25 years, several investigators tried to use machine learning or artificial neural network tools to generate a better prediction calculator [34–37]. All those attempts produced fair predictors but revealed the main obstacle in reaching such robust tool in IVF—we lack big databases. On one hand, some of the investigators claim on dozens of parameters that may influence live birth. However, most of the studies that have such a diversity of parameters included relatively small number of cycles. In summary, our ability to produce a robust live birth predictor in IVF will necessitate a multicenter cooperation to produce a big database with multiple cycles and diverse parameters.

Conclusion

AI is already impacting many areas in the industry and medicine and is likely to evolve, expand, and play a more dominant role in the reproductive field soon. It offers a more objective, accurate, and rapid data analysis. AI presents a more precise and individualized medicine based on algorithms integrating numerous variables, thus offering additive advantages to human capabilities. Applying this technology will enhance a reproducible, efficient, less biased, and more comprehensive decision-making both in the IVF lab and in the clinical field, thus improving IVF outcomes.

References

1. Camacho DM, Collins KM, Powers RK, Costello JC, Collins JJ. Next-generation machine learning for biological networks. Cell. 2018;173(7):1581–92.
2. Letterie G, MacDonald A. Artificial intelligence in in vitro fertilization: a computer decision support system for day-to-day management of ovarian stimulation during in vitro fertilization. Fertil Steril. 2020;114(5):1026–31.
3. Manna C, Nanni L, Lumini A, Pappalardo S. Artificial intelligence techniques for embryo and oocyte classification. Reprod Biomed Online. 2013;26(1):42–9.

4. Faghih RT, Styer AK, Brown EN. Automated ovarian follicular monitoring: a novel real-time approach. Annu Int Conf IEEE Eng Med Biol Soc. 2017;2017:632–5.
5. Bian N, Eramian MG, Pierson RA. Evaluation of texture features for analysis of ovarian follicular development. Med Image Comput Comput Assist Interv. 2006;9(Pt 2):93–100.
6. Hu SY, Xu H, Li Q, Telfer BA, Brattain LJ, Samir AE. Deep learning-based automatic endometrium segmentation and thickness measurement for 2D transvaginal ultrasound. Annu Int Conf IEEE Eng Med Biol Soc. 2019;2019:993–7.
7. Cavalera F, Zanoni M, Merico V, Bui TTH, Belli M, Fassina L, et al. A neural network-based identification of developmentally competent or incompetent mouse fully-grown oocytes. J Vis Exp. 2018;133:e56668.
8. Targosz A, Przystalka P, Wiaderkiewicz R, Mrugacz G. Semantic segmentation of human oocyte images using deep neural networks. Biomed Eng Online. 2021;20(1):40.
9. Javadi S, Mirroshandel SA. A novel deep learning method for automatic assessment of human sperm images. Comput Biol Med. 2019;109:182–94.
10. McCallum C, Riordon J, Wang Y, Kong T, You JB, Sanner S, et al. Deep learning-based selection of human sperm with high DNA integrity. Commun Biol. 2019;2:250.
11. Mirsky SK, Barnea I, Levi M, Greenspan H, Shaked NT. Automated analysis of individual sperm cells using stain-free interferometric phase microscopy and machine learning. Cytometry A. 2017;91(9):893–900.
12. Goodson SG, White S, Stevans AM, Bhat S, Kao CY, Jaworski S, et al. CASAnova: a multiclass support vector machine model for the classification of human sperm motility patterns. Biol Reprod. 2017;97(5):698–708.
13. Kanakasabapathy MK, Sadasivam M, Singh A, Preston C, Thirumalaraju P, Venkataraman M, et al. An automated smartphone-based diagnostic assay for point-of-care semen analysis. Sci Transl Med. 2017;9(382):eaai7863.
14. Dimitriadis I, Bormann CL, Kanakasabapathy MK, Thirumalaraju P, Kandula H, Yogesh V, et al. Automated smartphone-based system for measuring sperm viability, DNA fragmentation, and hyaluronic binding assay score. PLoS One. 2019;14(3):e0212562.
15. Storr A, Venetis CA, Cooke S, Kilani S, Ledger W. Inter-observer and intra-observer agreement between embryologists during selection of a single day 5 embryo for transfer: a multicenter study. Hum Reprod. 2017;32(2):307–14.
16. Hsu MI, Mayer J, Aronshon M, Lanzendorf S, Muasher S, Kolm P, et al. Embryo implantation in in vitro fertilization and intracytoplasmic sperm injection: impact of cleavage status, morphology grade, and number of embryos transferred. Fertil Steril. 1999;72(4):679–85.
17. Motato Y, de los Santos MJ, Escriba MJ, Ruiz BA, Remohi J, Meseguer M. Morphokinetic analysis and embryonic prediction for blastocyst formation through an integrated time-lapse system. Fertil Steril. 2016;105(2):376–84 e9.
18. Dirvanauskas D, Maskeliunas R, Raudonis V, Damasevicius R. Embryo development stage prediction algorithm for automated time lapse incubators. Comput Methods Prog Biomed. 2019;177:161–74.
19. Raudonis V, Paulauskaite-Taraseviciene A, Sutiene K, Jonaitis D. Towards the automation of early-stage human embryo development detection. Biomed Eng Online. 2019;18(1):120.
20. Thirumalaraju P, Kanakasabapathy MK, Bormann CL, Gupta R, Pooniwala R, Kandula H, et al. Evaluation of deep convolutional neural networks in classifying human embryo images based on their morphological quality. Heliyon. 2021;7(2):e06298.
21. VerMilyea M, Hall JMM, Diakiw SM, Johnston A, Nguyen T, Perugini D, et al. Development of an artificial intelligence-based assessment model for prediction of embryo viability using static images captured by optical light microscopy during IVF. Hum Reprod. 2020;35(4):770–84.
22. Tran D, Cooke S, Illingworth PJ, Gardner DK. Deep learning as a predictive tool for fetal heart pregnancy following time-lapse incubation and blastocyst transfer. Hum Reprod. 2019;34(6):1011–8.
23. Reignier A, Lammers J, Barriere P, Freour T. Can time-lapse parameters predict embryo ploidy? A systematic review. Reprod Biomed Online. 2018;36(4):380–7.

24. Zaninovic N, Irani M, Meseguer M. Assessment of embryo morphology and developmental dynamics by time-lapse microscopy: is there a relation to implantation and ploidy? Fertil Steril. 2017;108(5):722–9.
25. Chavez-Badiola A, Flores-Saiffe Farias A, Mendizabal-Ruiz G, Garcia-Sanchez R, Drakeley AJ, Garcia-Sandoval JP. Predicting pregnancy test results after embryo transfer by image feature extraction and analysis using machine learning. Sci Rep. 2020;10(1):4394.
26. Huang B, Tan W, Li Z, Jin L. An artificial intelligence model (euploid prediction algorithm) can predict embryo ploidy status based on time-lapse data. Reprod Biol Endocrinol. 2021;19(1):185.
27. Matson PL. Internal quality control and external quality assurance in the IVF laboratory. Hum Reprod. 1998;13(Suppl 4):156–65.
28. Franco JG Jr, Petersen CG, Mauri AL, Vagnini LD, Renzi A, Petersen B, et al. Key performance indicators score (KPIs-score) based on clinical and laboratorial parameters can establish benchmarks for internal quality control in an ART program. JBRA Assist Reprod. 2017;21(2):61–6.
29. Castillo CM, Harper J, Roberts SA, O'Neill HC, Johnstone ED, Brison DR. The impact of selected embryo culture conditions on ART treatment cycle outcomes: a UK national study. Hum Reprod Open. 2020;2020(1):hoz031.
30. Embryology ESIGO, Alpha Scientists in Reproductive Medicine. The Vienna consensus: report of an expert meeting on the development of ART laboratory performance indicators. Reprod Biomed Online. 2017;35(5):494–510.
31. Bormann CL, Curchoe CL, Thirumalaraju P, Kanakasabapathy MK, Gupta R, Pooniwala R, et al. Deep learning early warning system for embryo culture conditions and embryologist performance in the ART laboratory. J Assist Reprod Genet. 2021;38(7):1641–6.
32. McLernon DJ, Steyerberg EW, Te Velde ER, Lee AJ, Bhattacharya S. Predicting the chances of a live birth after one or more complete cycles of in vitro fertilisation: population based study of linked cycle data from 113,873 women. BMJ. 2016;355:i5735.
33. Ratna MB, Bhattacharya S, Abdulrahim B, McLernon DJ. A systematic review of the quality of clinical prediction models in in vitro fertilisation. Hum Reprod. 2020;35(1):100–16.
34. Kaufmann SJ, Eastaugh JL, Snowden S, Smye SW, Sharma V. The application of neural networks in predicting the outcome of in-vitro fertilization. Hum Reprod. 1997;12(7):1454–7.
35. Banerjee P, Choi B, Shahine LK, Jun SH, O'Leary K, Lathi RB, et al. Deep phenotyping to predict live birth outcomes in in vitro fertilization. Proc Natl Acad Sci U S A. 2010;107(31):13570–5.
36. Vogiatzi P, Pouliakis A, Siristatidis C. An artificial neural network for the prediction of assisted reproduction outcome. J Assist Reprod Genet. 2019;36(7):1441–8.
37. Choi B, Bosch E, Lannon BM, Leveille MC, Wong WH, Leader A, et al. Personalized prediction of first-cycle in vitro fertilization success. Fertil Steril. 2013;99(7):1905–11.

15 PGT-A also Known as PGS: The Indications

Andreas G. Schmutzler

Definitions

This chapter is dealing with the view of the physician, i.e. gynaecologist and specialist in reproductive medicine, on the genetic detection of presumed anomalies of chromosomes of oocytes and embryos as part of a treatment with artificial reproductive technologies (ART), i.e. IVF or ICSI. This in turn is done in the broadest sense as a treatment of sterility. The analysis is done after oocyte retrieval and before embryo transfer, for the detection of numerical pathologies, i.e. aneuploidies.

There are two important subgroups: the first subgroup, where one or both of the future parents are suffering from a hereditary disease or are carrier thereof, and the second subgroup, where both are healthy in this regard.

Historically, a distinction between these subgroups has been made by the terms preimplantation genetic diagnosis (PGD) for the first and preimplantation genetic screening (PGS) for the second.

To be more precise, PGD describes a case group in which the indication for the investigation is "hereditary disease in future parents", in order to avoid this in the offspring. And PGS describes a case group in which the indication is "suspected genetic disorders at the level of gametes and embryos", in order to increase the success rates of the treatment with in vitro fertilization.

This chapter is a newly edited, updated, partly shortened, partly extended textbook version of a journal article of the author [1].

A. G. Schmutzler (✉)
University Women's Hospital, Christian-Albrechts-University, Kiel, Germany

gyn-medicum, Center for Reproductive Medicine, Göttingen, Germany
e-mail: schmutzler@email.uni-kiel.de

J. G. Schenker et al. (eds.), *Hot Topics in Human Reproduction*, Reproductive Medicine for Clinicians 3, https://doi.org/10.1007/978-3-031-24903-7_15

These success rates could be, from the perspective of the five main professionals involved

- Geneticist: quick, low-cost, simple, few investigations with precise reliable genetic results.
- Embryologist: quick, low-cost, simple treatments of few oocytes, sperms and embryos with a high rate of implantation followed by the birth of a healthy singleton child at term.
- Reproductive endocrinologist: short time to a healthy singleton child, few simple low-cost treatments with a low rate of complications, miscarriages and psychological stress to the patients.
- Obstetrician: healthy clinical and ongoing pregnancy, few non-invasive investigations, healthy spontaneous delivery at term.
- Neonatologist: healthy mature child with normal birth weight.

Historically, one would either try to detect a monogenetic disorder in one or two biopsied blastomeres from an eight-cell day 3 "embryo" (more precise pre-embryo, as trophoblast and embryoblast morphologically—if at all—are not yet distinguishable) by PCR. Or, one would try to detect an aneuploidy in five or some more chromosomes by FISH, but not both at the same time!

Nowadays, for the first subgroup ("hereditary") one could, in about five trophoblast cells of a day 5 or 6 blastocyst on the one hand, look for a monogenetic disorder and at the same time check for aneuploidies or, vice versa in a "non-hereditary" case, check for aneuploidies and at the same time for a panel of genes, e.g. for a carrier status, currently by NGS. Both approaches might be contested for ethical reasons. But it might be difficult to defend not to look for aneuploidies, when looking for monogenetic defects.

Finally, it is thinkable that in the future in both subgroups, one might look, when checking the health of the embryo in vitro, not only for aneuploidies but also for any other genetic or epigenetic defects, e.g. single genes steering the development in vitro or in vivo.

Now, the terminology in this field has recently been "officially" changed: PGD was changed to "structural rearrangement testing (PGT-SR)" and "monogenetic disorder testing (PGT-M)", and PGS was changed to "aneuploidy testing (PGT-A)". This classification takes the view of the laboratory and not of the gynaecologist treating the patients. It is focusing on the methods applied in the genetic laboratory and not on the medical indication to do so.

So, there are several considerations: the term "PGT-A" might get us to believe that this method is only indicated when treating non-hereditary cases. The term "PGT-SR" and "PGT-M" might make us believe that "PGT-A" is not indicated. The medical indication for any of these is not visible. The terms "screening" and "test" suggest different settings. In a screening the indication is not a pathological finding per se but risk factors, and one has to scan many in order to find a few pathologies. In a test the indication might or might not be a pathologic finding, and one expects a much higher yield. So, in this last regard, the change from screening as PGS to

PGT-A seems to be justified, as, dependent on age, in average at least half of all oocytes are aneuploid. Finally, probable future developments are not mirrored in that scheme: testing in parallel aneuploidy, single genes and other issues, like epigenetics, proteomics and metabolomics, to detect the developmental competence of embryos.

For the physicians' decision to advise for a screening or a test, there are two major considerations: reliability of the diagnostic method and its indication. In the field of PGT primarily the method gets discussed, by the scientists, whereas the normally also important indication, not, may be due to the often not present physicians. In specialized meetings in the field, they are in the vast minority.

As many of the PGT tests are reliable now, the primary medical focus now should be laid on the indication. This is even more true, as the methods were and are—fortunately—pushed by scientists in the lab and unfortunately in the beginning, and still in many countries, mostly ignored by the physicians. But also, the history of studies in PGT-A points in this direction: some primarily designed with a focus on the laboratory evaluation of the method, and thus sometimes ignoring indications, led to sometimes wrong clinical conclusions; see below.

For this reason, for discussing indications, one has to distinguish the two major case groups, event-related case groups, as is done in this report: if the cause of the investigation lies in a disease in the family, PGD could be used, done by one or some of the PGTs (-SR, -M and -A). Otherwise, at the moment for the lack of anything better, PGS is also still being used by IVF registries.

One argument to use only PGT could be that ultimately there is no such thing as complete genetic normality in any human being. So, it would be discussed to test for a panel of frequent and less frequent deviations. And it will be discussed, if some would signify just a quantitative or also a qualitative difference. But also there, one should concede that in a complete discussion of the justification of any investigation, the justification should be not solely focused on the scientific methodology but also on the medical indication.

In general, PGT in routine clinical practice is understood to be invasive diagnostics with a biopsy of polar bodies of oocytes, blastomeres of eight-cell embryos and trophoblast cells of blastocysts, followed by indication-dependent relevant genetic analysis, as performed in this report. Experimentally, "semi-invasive" (aspiration of blastocoel fluid) and non-invasive (analysis of the culture medium) methods have been proposed. These genetic and embryology lab aspects of PGT will not be evaluated here and also not the indications for PGT-SR and PGT-M. The focus of this chapter is on the primary medical point of view, the indication of PGT-A also known as PGS.

Indications

First, a distinction must be made between goals and indications. In contrast to popular belief, the "pregnancy rate" is not the only aim. Instead, there are five distinct aims that partly compete with one another [2]. The possible aims could be to:

- Increase the pregnancy rate.
- Reduce the miscarriage rate.
- Reduce the multiple birth rate.
- Reduce the malformation rate.
- Reduce the rate of pointless treatments with artificial reproductive technology (ART, i.e. IVF or ICSI).

Furthermore, various indications were discussed, such as:

- Advanced maternal age (AMA).
- Repeated implantation failure (RIF).
- Repeated miscarriage (RM).
- Severe male factor (SMF) infertility.

The basic idea is that if one tries to implant only euploid embryos into the uterus, one can improve the patient's situation. However, experience and general principles show that this is not that easy [2].

Pregnancy

Experience with the eight-cell embryo biopsy showed that the intervention had a negative effect on pregnancy rates to some extent. For this reason, this approach was ultimately abandoned for PGS after a long dispute. To lessen the trauma to the embryo, blastocysts are being biopsied. Only in countries where this is legally problematic, oocytes are biopsied for analysis of polar bodies. Based on this experience, hardly anyone claims that there are no effects of these biopsies on embryo development. Thus, if the primary goal is to improve the pregnancy rate, PGT-A makes sense if a stochastic selective benefit is to be expected.

We can consider different settings: the indication "to increase the chance of pregnancy" is there, if the probability to transfer an euploid embryo theoretically might be increased. The real benefit of the test depends in a retrospective view on the outcome of the test.

> "The *normal* case": we receive ten oocytes. Six of them are fertilized, and three develop to blastocysts on day 5, two of which are euploid. If one only intends to transfer one blastocyst into the uterus, the selection advantage can be calculated as 100% euploid after diagnosis and a 67% chance of a euploid blastocyst without PGS, so there is a 50% selection advantage (from 67 to 100%). This advantage will most likely outweigh the disadvantage of biopsy trauma. There is an indication and a benefit to do PGT-A.

> "The case of *good genetic* embryo quality": all three embryos are euploid. Then, the selection advantage in terms of the pregnancy rate is zero. An indication was there, but retrospectively no benefit.

There is no indication "pregnancy increase", if the probability to transfer an euploid embryo definitely cannot be increased.

"The case of *poor development*": only one blastocyst develops. Then, the selection advantage is zero, and the primary goal of increasing the pregnancy rate may be slightly endangered. There is no indication to do PGT-A.

But there is an indication "pregnancy increase", if a fast success is desired. So, other factors influencing the chance of pregnancy must be considered. After the above, the chance of pregnancy with the first fresh transfer may be increased by PGT-A. However, considering the chance independent of time and adding the odds of fresh and further transfers after cryopreservation (cumulative pregnancy rate), this chance without PGT-A could be higher than that of a single fresh PGS transfer. This idea is highly controversial. In younger patients with good blastocyst morphology, PGT-A could not improve the cumulative pregnancy rate ([3], s. Table 15.1).

On the one hand, the blastocyst culture per se could be disadvantageous if a longer culture time of 5 days instead of 2–3 days in vitro is worse than the "blastocyst

Table 15.1 RCTs with PGT-A

First author	Methods	Clinical results
Schoolcraft et al. [4]	>35 years, blastocyst biopsy, aSNP	Implantation increased (71% vs. 46%)
Yang et al. [5]	32 years, blastocyst biopsy, aCGH	Pregnancy per embryo transfer increased (71% vs. 46%)
Forman et al. [6]	35 years, blastocyst biopsy, single embryo transfer with PGS vs. double embryo transfer without PGT-A, RT-PCR	Multiples reduced (0% vs. 65%), pregnancy same (61% vs. 65%)
Rubio et al. [7]	43 years, biopsy of eight-cell embryo, FISH	Birth per cycle increased (24% vs. 11%)
Scott Jr et al. [8]	32 years, blastocyst biopsy, RT-PCR	Birth per cycle increased (85% vs. 68%)
Chen et al. [9]	7 trials (including 4 RCTs)	Implantation, clinical pregnancy, ongoing pregnancy, live birth increased; miscarriage, multiples reduced
Dahdouh et al. [10]	8 trials (including 3 RCTs)	Implantation increased
Verpoest et al. [11]	36–40 years, polar bodies, aCGH	Implantation increased, miscarriages decreased, less interventions
Munné et al. [12]	25–40 years, frozen-thawed SET, NGS	35–40 years ongoing pregnancy per transfer increased
Simopoulou et al. [13]	11 RCTs	>35 years live birth increased
Yan et al. [3]	20–37 years, ≥3 good blastocysts, SET	Miscarriage decreased, cumulated live birth not better
Shi et al. [14]	9 RCTs, AMA	Live birth increased

RCT randomized controlled trial, *PGT-A* preimplantation genetic testing for aneuploidies, *aSNP* array single nucleotide polymorphism, *RT-PCR* real-time polymerase chain reaction, *FISH* fluorescence in situ hybridization, *aCGH* array comparative genome hybridization, *SET* single embryo transfer, *NGS* next-generation sequencing, *AMA* advanced maternal age culture in vivo" after transfer on day 2 or 3 after oocyte retrieval. On the other hand, the transfer of blastocysts on day 5, the day on which implantation takes place physiologically, might improve implantation.

"The case of *fast success*": the patient, who is 36 years old, has many oocytes. For her, it is more important to have a higher chance of success in the first transfer than to take the time to "blindly" undergo a fresh transfer first and cryo-transfers later on. Then, there is an indication to do PGT-A.

Miscarriage

It is known that the rate of miscarriages increases with age and that the dominant cause is the aneuploidy of embryos. The increase in the aneuploidy rate of the oocytes matches with increasing age. Similarly, the pregnancy rate drops drastically after 40 years. There are two different cases that can be considered.

There is an indication "decrease of miscarriage" after several miscarriages. Even if the patient has a chance to get a child after zero, one or several further miscarriages, the risk of miscarriage is decreased for the next pregnancy if done with PGT-A. The recommendation of the German Society of Gynaecology and Obstetrics to not apply PGT-A in these cases because of a long-time positive prognosis of recurrent miscarriages appears to be cruel.

"The case of *recurrent miscarriages*": the patient, who is 37 years old, has one child, has a normal ovarian reserve and has had three miscarriages. She wants a second child but, even more importantly, primarily no further miscarriages. There is an indication for PGT-A.

There is an indication "decrease of miscarriage" also in cases where the risk of a miscarriage because of elevated maternal age is high, even if not yet realized, also in combination with the indication "pregnancy increase" in order to shorten time to pregnancy. A sterility treatment with untested embryos might lead to several embryo transfers, fresh and frozen-thawed, plus miscarriages with curettage and waiting time before starting a new therapy—at a precious time when the pregnancy chance comes to its end.

"The case of *advanced maternal age*": the patient, who is 41 years old, has had no pregnancies, has a slightly reduced ovarian reserve, and is afraid of taking too much time, especially due to miscarriages and the associated loss of time endangering her likelihood of having children. There is an indication for PGT-A.

Malformation

If the exploration of the patient's preference shows that her primary goal is to reduce the risk of another abortion with a medical indication, then it makes sense to analyse a single existing blastocyst. This goal thus competes with the pregnancy chance. This is the textbook example that the indication of PGT-A needs to be detected by exploration of the will of the patient by the treating physician, i.e. the specialist in reproductive medicine, and cannot be detected theoretically or be done by a decision at the "green table" in the lab.

The example is the "case of *trauma interruption*": the patient is 40 years old, has no children, has a reduced ovarian reserve and has had a medically induced termination of one pregnancy due to trisomy 21. The doctor recommends that the biopsy of the only embryo does not take place for the sake of safety so as not to jeopardize the chance of pregnancy. The patient explains after the conflicting goals are clarified: reducing the risk of re-interruption is more important to her than increasing the chance of pregnancy. So there is an indication for PGT-A.

Multiples

Similarly, an exploration may indicate that the patient does not want to have multiple children per birth, also at the risk of reducing the chance of pregnancy.

This is the "case of being *afraid of multiples*": the patient, who is 38 years old, has three children from her first marriage, including twins, and two blastocysts have developed; however, she does not want to have both transferred. At the same time, she wants to increase the chance of quick success. When both embryos are euploid, the selection advantage is zero, and one of the embryos would be frozen. There is an indication for PGT-A, but no benefit.

If only one is euploid, the test also stochastically makes sense. So, there would be an indication for PGT-A and additionally also a benefit in terms of an increase of the pregnancy chance.

If both are aneuploid, the treatment would be shortened because no cryopreservation and second transfer would be performed. So, there is an indication for PGT-A and additionally a benefit by shortening the time to pregnancy.

Pointless ART Treatment

Finally, certain patients may be at an increased risk for an unusually low rate of euploid oocytes and embryos. The expectation values are approximately 50% for patients under 35 years of age, 33% for patients between 35 and 40 years of age, 25% for patients who are 40 years of age and below 25% for patients over 40 years of age.

PGS for this indication converts a therapeutic procedure, IVF, to a diagnostic procedure.

The example is the "case of *many treatments*": the patient, who is 32 years old, has had no pregnancies, had three oocyte retrievals and had six embryo transfers, which were fresh or cryo-transfers. She wants to know if continuation of therapy makes sense. Polar body biopsy results of the first PGT-A reveal that nine of ten oocytes are aneuploid, and the euploid egg did not develop into a blastocyst. When PGT-A is repeated, all eight oocytes are aneuploid. The patient opts for egg donation. So, there is an indication for PGT-A.

Advanced Maternal Age (AMA), Repeated Miscarriage (RM), Repeated Implantation Failure (RIF) and Severe Male Factor (SMF)

There are positive findings for AMA (implantation, ongoing pregnancy, birth, live birth, miscarriage and interventions) and also several for RM (both see Table 15.1). These findings must be combined with the need of an individual indication as described above. There are no RCTs for RIF and SMF yet. As always, all study findings are a basis for the treating physician for the decision, common with the couple, about the presence of an individual indication, i.e. make it more or less likely.

Ethics

Evidence-Based Medicine (EBM)

Evidence-based medicine distinguishes three levels

- "Top" (level I) randomized controlled trials (RCTs) and their meta-analyses
- "Centre" (level II) controlled, cohort or case-control studies
- "Bottom" (level III) estimations of authorities based on experience or first principles

When exploring new approaches, for ethical or logical reasons, the approach from bottom to top must be followed. When first principles, such as mathematics, e.g. stochastics, do not allow an advantage of a method, it is pointless and unethical to conduct further studies on this. If there is no single study that has provided proof of a principle to date, it is unethical to randomize patients to prove this principle.

Observational studies are usually conducted with "favourable cases", i.e. with patients for whom a benefit appears most likely. The patients who are individually selected for the purpose of a healing attempt are usually in a serious situation, and there is a suspicion of the chance of a cure by the new treatment method (diagnosis or therapy). These studies must have an ambitious goal, i.e. a high benefit because if the benefit is low, it is to be expected that when widely used, the benefit will disappear.

If the results of these studies have made the effectiveness of the method likely in terms of "proof of principle", it is ethically possible to randomize large groups of patients. On a broader basis, it must be determined whether the method only works for selected cases in the hands of a few specialists or for a large case group with many different practitioners. Only then the method can be recommended to the general public outside of studies for proven indications. For this reason, the goal of such a study may be significantly smaller than that of a study of the principle because small improvements are usually clinically important.

Design of Studies

These considerations are significant in the design of a level I (RCT) study. On the one hand, one should not withhold even small advances from the general population. On the other hand, the smaller the progress being studied, the more complex the investigation will be; that is, the investigation will be more expensive and time-consuming.

Statistically, more patients are needed to prove small differences, which increases costs. Additionally, if necessary, the duration is extended, as more patients must be recruited. Both can lead to the investigation not being carried out, either because the study is too expensive or pointless because one can expect that newer methods will be introduced after the investigation has ended.

Reflecting this, the decision not to investigate would possibly hurt less than the decision to perform a study that restricts the number of patients only due to a lack of financial resources and thus sets very high targets, in order to correctly claim that the high goal was not achieved. At the same time, however, there is a great danger that, because of the high quality of a level I study, the audience will draw the wrong conclusion that the method is ineffective. This may deprive the general public of a minor but clinically significant advance.

Design of PGT-A Studies

For PGS trials, this means that we have to distinguish two stages of the investigations

- The first is the "proof of principle". If one intends to investigate if PGS works at all, the numerator and denominator in the cascade must be close to one other, preferably the number of biochemical pregnancies to the number of embryo transfers or the number of transferred embryos to the number of implanted embryos. The closer the examination points are to each other, the lower the number of cases will be needed for the detection of statistically significant differences, and the smaller the detectable differences will be between the PGT group and the non-PGT group as the control.
- The second is the "efficacy study". If one intends to determine if a large group of patients benefits from the care of multiple physicians and in multiple settings, one should use an RCT with the starting point "intention to treat" (ITT). However, the point at which randomization occurs is most important. It makes no sense to use the first contact as the starting point and the birth of a healthy child as the endpoint because other factors, such as financial costs, might play a greater role than the effectiveness of PGT.

Likewise, the use of the start of ovarian stimulation as a starting point is not indicated because at that time, it is still unclear how many oocytes, fertilizations, embryos and blastocysts will be present. If this is disregarded, there is a risk that biopsy will occur according to the protocol, without a stochastic selection advantage being present. This may result in a reduction rather than an increase in the pregnancy rate, and the method would be discredited falsely.

Therefore, the patient must be informed of the method twice, namely, at the start of the stimulation and immediately before the biopsy, to consider the possible advantages or disadvantages. Thus, an embryo biopsy according to the protocol, with no consideration of the number of embryos and the desired number of embryos to be transferred and with the sole aim of increasing the pregnancy rate, appears to be unethical.

PGT-A Studies

The origin of PGS of human embryos is based on human PGD [15]. Subsequently, the method has been extrapolated for the screening of oocytes [16, 17] and embryos. In oocytes, the first and second polar bodies, in embryos, from one to two cells of an eight-cell embryo, are examined by FISH with five to nine probes.

The above-mentioned goals and indications were developed until approximately 2010. Numerous studies (EBM levels II and I) have been performed to attempt to prove the effectiveness of the method. The goal of increasing the pregnancy rate could not be demonstrated, although more than ten level I studies were also carried out for this purpose (see [18]).

If the study design did not adequately align the biopsy with the stochastic criteria (see [19]), the pregnancy rate was, as expected, even lower. The goal of reducing the miscarriage rate has been pursued since 1999 in numerous publications of level II studies by Munne et al. [20] but has often received little attention in discussions. With the use of FISH, however, the notion that at least about half of the oocytes in humans are aneuploid has been undisputed.

Thus, the reason for the lack of success of the methodology has been unclear. Unusually, due to the importance of the issue, the largest European professional society in the field, namely, the European Society of Human Reproduction and Embryology (ESHRE), decided to solve this puzzle by sponsoring studies, together with the company Blue Gnome, later bought by Illumina. This new approach, by ESHRE and others, was later called the onset of "PGS 2.0".

The effectiveness of the method should be increased by applying strict standard operation procedures [21], reducing trauma and increasing the analysis, i.e. by advancing the biopsy to the oocyte and analysing all chromosomes with aCGH. The pilot study showed the high effectiveness of the chips. It also showed that in 40-year-old women, on average, only one in four oocytes is euploid [22]. Subsequently, an international multicentre RCT was launched to investigate the increase in the pregnancy rates in AMA. The available resources allowed the randomization of 600 patients, resulting in a 15%-point study goal of increasing the pregnancy rates.

At the same time, beginning in 2012, the first RCTs appeared, which also used comprehensive chromosome screening (CCS) and, to reduce trauma, postponed the biopsy to the blastocyst stage. Only now, for the first time, all RCTs found significant advantages for PGS (Table 15.1).

After recruitment delays, the results of the ESHRE study were published in 2018 [11]. The study goal was not achieved, but it showed that the implantation rate was increased by PGS.

PGT-A: Use and Opinions

Use

The method is increasingly used around the world (compare to [23]), especially in the USA. Last data show that in Europe, it is done in 4% percent of all ART treatments [24], in the USA in 44% ([25] for 2019), in Australia in 13% (ANZARD, [26]) and globally in 4% (ICMART, [27]).

Opinions

There is still disagreement about the interpretation of the PGS 2.0 results. However, when examining the opinion publications regarding this purpose, one finds that it is striking that the effectiveness is predominantly assumed:

- In an international survey study, the majority opinion was that PGS is evidence-based medicine, increases live birth rates, reduces miscarriage rates and should be performed with an indication, primarily for repeated implantation failure, in less than 20% of the cycles (IVF-Worldwide Survey, [28]).
- The "Virtual Academy of Genetics" stated that PGS is not experimental, increases live birth rates and reduces miscarriage and multiple birth rates [29].
- In an expert's opinion paper, the majority thinks that PGS increases live birth rates and reduces the time to pregnancy [30].
- The forum COGEN (Controversies in Genetics, a Series of international congresses) stated that PGS is evidence-based medicine and that a pragmatic approach is favoured [31].
- The authors of the ESHRE RCT found that the results "point to a clinical benefit".
- The American Society of Reproductive Medicine [32] stated that PGS "will likely be part of a future multidimensional approach", but it does not recommend "routine use of blastocyst biopsy with aneuploidy testing in all infertile patients". This is in accordance with the indications examined here.

When does a physician change his current treatment routine? Presumably, when a meta-analysis with enough RCTs suggests that another approach is more successful or when the vast majority of his colleagues change their treatment approaches. Thus far, globally this is not yet the case with PGT-A.

However, every physician specialist in reproductive medicine must address this topic. Ultimately, an ethically justifiable decision does not require a meta-analysis, an RCT or any other trial. However, several RCTs indicate that it is likely that PGT-A, if strictly indicated, while taking into account stochastics and patient preference, can benefit the patient.

Therefore, the task of treatment specialists in reproductive medicine is similar to the counselling of specialists in prenatal medicine: present all methods of investigation of the embryo and respect the patient's preference.

Acknowledgements I thank my wife and clinical partner Dr. med. Monica Tobler, specialist in reproductive medicine, for many clinical and scientific discussions and the critical review of this chapter.

Declaration Because this investigation analysed theories and the literature and because new patient examinations were not conducted, ethical approval was not required.

Funding This study was funded by gyn-medicum, Center for Reproductive Medicine, Goettingen, Germany.

References

1. Schmutzler AG. Theory and practice of preimplantation genetic screening (PGS). Eur J Med Genet. 2019;62(8):103670.
2. Schmutzler AG, Filges I, Al-Hasani S, Diedrich K, Miny P. The future of aneuploidy screening. Diagnosis first! Gynäkologe. 2014a;47(4):263–70.
3. Yan J, Qin Y, Zhao H, Sun Y, Gong F, Li R, Sun X, Ling X, Li H, Hao C, Tan J, Yang J, Zhu Y, Liu F, Chen D, Wei D, Lu J, Ni T, Zhou W, Wu K, Gao Y, Shi Y, Lu Y, Zhang T, Wu W, Ma X, Ma H, Fu J, Zhang J, Meng Q, Zhang H, Legro RS, Chen ZJ. Live birth with or without preimplantation genetic testing for aneuploidy. N Engl J Med. 2021;385(22):2047–58.
4. Schoolcraft WB, Surrey E, Minjarez D, Gustofson RL, Scott RT Jr, Katz-Jaffe MG. Comprehensive chromosome screening (CCS) with vitrification results in improved clinical outcome in women >35 years: a randomized control trial. Fertil Steril. 2012;98(3, Supplement):1.
5. Yang Z, Liu J, Collins GS, Salem SA, Liu X, Lyle SS, Peck AC, Sills ES, Salem RD. Selection of single blastocysts for fresh transfer via standard morphology assessment alone and with array CGH for good prognosis IVF patients: results from a randomized pilot study. Mol Cytogenet. 2012;5:24.
6. Forman EJ, Hong KH, Ferry KM, Tao X, Taylor D, Levy B, Treff NR, Scott RT Jr. In vitro fertilization with single euploid blastocyst transfer: a randomized controlled trial. Fertil Steril. 2013;100(1):100–7.
7. Rubio C, Bellver J, Rodrigo L, Bosch E, Mercader A, Vidal C, De los Santos MJ, Giles J, Labarta E, Domingo J, Crespo J, Remohí J, Pellicer A, Simón C. Preimplantation genetic screening using fluorescence in situ hybridization in patients with repetitive implantation failure and advanced maternal age: two randomized trials. Fertil Steril. 2013;99:1400–7.
8. Scott RT Jr, Upham KM, Forman EJ, Hong KH, Scott KL, Taylor D, Tao X, Treff NR. Blastocyst biopsy with comprehensive chromosome screening and fresh embryo transfer significantly increases in vitro fertilization implantation and delivery rates: a randomized controlled trial. Fertil Steril. 2013;100(3):697–703.

9. Chen M, Wei S, Hu J, Quan S. Can comprehensive chromosome screening technology improve IVF/ICSI outcomes? A meta-analysis. PLoS One. 2015;10(10):e0140779.
10. Dahdouh EM, Balayla J, García-Velasco JA. Comprehensive chromosome screening improves embryo selection: a meta-analysis. Fertil Steril. 2015;104(6):1503–12.
11. Verpoest W, Staessen C, Bossuyt PM, Goossens V, Altarescu G, Bonduelle M, Devesa M, Eldar-Geva T, Gianaroli L, Griesinger G, Kakourou G, Kokkali G, Liebenthron J, Magli MC, Parriego M, Schmutzler AG, Tobler M, van der Ven K, Geraedts J, Sermon K. Preimplantation genetic testing for aneuploidy by microarray analysis of polar bodies in advanced maternal age: a randomized clinical trial. Hum Reprod. 2018;33(9):1767–76.
12. Munné S, Kaplan B, Frattarelli JL, Child T, Nakhuda G, Shamma FN, Silverberg K, Kalista T, Handyside AH, Katz-Jaffe M, Wells D, Gordon T, Stock-Myer S, Willman S, STAR Study Group. Preimplantation genetic testing for aneuploidy versus morphology as selection criteria for single frozen-thawed embryo transfer in good-prognosis patients: a multicenter randomized clinical trial. Fertil Steril. 2019;112(6):1071–1079.e7.
13. Simopoulou M, Sfakianoudis K, Maziotis E, Tsioulou P, Grigoriadis S, Rapani A, Giannelou P, Asimakopoulou M, Kokkali G, Pantou A, Nikolettos K, Vlahos N, Pantos K. PGT-A: who and when? A systematic review and network meta-analysis of RCTs. J Assist Reprod Genet. 2021;38(8):1939–57.
14. Shi WH, Jiang ZR, Zhou ZY, Ye MJ, Qin NX, Huang HF, Chen SC, Xu CM. Different strategies of preimplantation genetic testing for aneuploidies in women of advanced maternal age: a systematic review and meta-analysis. J Clin Med. 2021;10(17):3895.
15. Handyside AH, Kontogianni EH, Hardy K, Winston RM. Pregnancies from biopsied human preimplantation embryos sexed by Y-specific DNA amplification. Nature. 1990;344:768–70.
16. Schmutzler AG, Schmutzler RK, Krebs D, Gordon JW. Polar body and blastomere biopsy in the mouse. In: Second joint meeting of the European Society of Human Reproduction and Embryology and the European Sterility Congress Organization, Milan, Italy. Human Reproductive Programme and Abstracts. Milan: European Sterility Congress Organization; 1990. p. 7–8.
17. Verlinsky Y, Ginsberg N, Lifchez A, Valle J, Moise J, Strom CM. Analysis of the first polar body: preconception genetic diagnosis. Hum Reprod. 1990;7:826–9.
18. Geraedts J, Collins J, Gianaroli L, Goossens V, Handyside A, Harper J, Montag M, Repping S, Schmutzler A. What next for preimplantation genetic screening? A polar body approach! Hum Reprod. 2010;25:575–7.
19. Mastenbroek S, Twisk M, van Echten-Arends J, Sikkema-Raddatz B, Korevaar JC, Verhoeve HR, Vogel NE, Arts EG, de Vries JW, Bossuyt PM, Buys CH, Heineman MJ, Repping S, van der Veen F. In vitro fertilization with preimplantation genetic screening. N Engl J Med. 2007;357(1):9–17.
20. Munne S, Magli C, Cohen J, Morton P, Sadowy S, Gianaroli L, Tucker M, Márquez C, Sable D, Ferraretti AP, Massey JB, Scott R. Positive outcome after preimplantation diagnosis of aneuploidy in human embryos. Hum Reprod. 1999;14:2191–9.
21. Harton G, Braude P, Lashwood A, Schmutzler A, Traeger-Synodinos J, Wilton L, Harper JC. ESHRE PGD consortium best practice guidelines for organization of a PGD centre for PGD/preimplantation genetic screening. Hum Reprod. 2011;26:14–24.
22. Geraedts J, Montag M, Magli MC, Repping S, Handyside A, Staessen C, Harper J, Schmutzler A, Collins J, Goossens V, van der Ven H, Vesela K, Gianaroli L. Polar body array CGH for prediction of the status of the corresponding oocyte. Part I: clinical results. Hum Reprod. 2011;26:3173–80.
23. Schmutzler AG, von Otte S, Tobler M, Filges I, Eckmann-Scholz C, Miny P. Global state of preimplantation genetic diagnosis. Frequency of application and indications. Gynäkologe. 2014;47(8):571–6.
24. European IVF-Monitoring Consortium (EIM) for the European Society of Human Reproduction and Embryology (ESHRE), Wyns C, De Geyter C, Calhaz-Jorge C, Kupka MS, Motrenko T, Smeenk J, Bergh C, Tandler-Schneider A, Rugescu IA, Vidakovic S, Goossens

V. ART in Europe 2017: results generated from European registries by ESHRE. Hum Reprod Open. 2021;3:hoab026.
25. Centers for Disease Control and Prevention. 2019 Assisted reproductive technology fertility clinic and national summary report. Atlanta: US Dept of Health and Human Services; 2021.
26. Newman JE, Paul RC, Chambers GM. Assisted reproductive technology in Australia and New Zealand 2018. Sydney: University of New South Wales; 2020.
27. Chambers GM, Dyer S, Zegers-Hochschild F, de Mouzon J, Ishihara O, Banker M, Mansour R, Kupka MS, Adamson GD. International Committee for Monitoring Assisted Reproductive Technologies world report: assisted reproductive technology, 2014. Hum Reprod. 2021;36(11):2921–34.
28. Weissman A, Shoham G, Shoham Z, Fishel S, Leong M, Yaron Y. Preimplantation genetic screening: results of a worldwide web-based survey. Reprod Biomed Online. 2017; 35(6):693–700.
29. Fiorentino F, Fishel S, Franasiak F, Gianaroli L, Gordon T, Griffin D, Grifo J, Hamamah S, Handyside A, Horowitz A, Hughes M, Leong M, Munne S, Nánássy L, Schmutzler A, Scott R, Shoham Z, Shulman L, Vereczkey A, Weissman A, Wells D, Yaron Y. A statement on the use of preimplantation genetic screening (PGS) of chromosomes for IVF patients. Paris: Virtual Academy of Genetics; 2015. https://ivf-worldwide.com
30. Sermon K, Capalbo A, Cohen J, Coonen E, De Rycke M, De Vos A, Delhanty J, Fiorentino F, Gleicher N, Griesinger G, Grifo J, Handyside A, Harper J, Kokkali G, Mastenbroek S, Meldrum D, Meseguer M, Montag M, Munné S, Rienzi L, Rubio C, Scott K, Scott R, Simon C, Swain J, Treff N, Ubaldi F, Vassena R, Vermeesch JR, Verpoest W, Wells D, Geraedts J. The why, the how and the when of PGS 2.0: current practices and expert opinions of fertility specialists, molecular biologists, and embryologists. Mol Hum Reprod. 2016;22(8):845–57.
31. Griffin DK, Fishel S, Gordon T, Yaron Y, Grifo J, Hourvitz A, Rechitsky S, Elson J, Blazek J, Fiorentino F, Treff N, Munne S, Leong M, Schmutzler A, Vereczkey A, Ghobara T, Nánássy L, Large M, Hamamah S, Anderson R, Gianaroli L, Wells D. Continuing to deliver: the evidence base for pre-implantation genetic screening. BMJ. 2017;14:356.
32. Practice Committees of the American Society for Reproductive Medicine and the Society for Assisted Reproductive Technology. The use of preimplantation genetic testing for aneuploidy (PGT-A): a committee opinion. Fertil Steril. 2018;109(3):429–36.

16 The Essential Role of In Vitro Maturation in Assisted Reproduction

Justin Tan and Seang Lin Tan

Introduction

Notwithstanding the burgeoning success of in vitro fertilization (IVF) [1], it is important to recognize the seminal role of in vitro maturation (IVM) in the development of IVF as well as its current widespread application amongst the arsenal of assisted reproductive techniques for over 25 years [2, 3]. While traditional IVF involves recovery of in vivo maturated oocytes, IVM refers to the recovery of immature oocytes from small antral follicles at the germinal vesicle (GV) or metaphase I (MI) stage and subsequent meiotic resumption under specifically controlled culture conditions. Cha et al. reported the first birth using IVM of immature oocytes collected at caesarean section within an oocyte donation programme in 1991 [4], but it was only after Trounson and colleagues reported the first pregnancy using a woman's own immature oocytes collected by transvaginal ultrasound-guided follicle aspiration that IVM emerged as a viable alternative to IVF for patients with polycystic ovaries [5]. The

J. Tan
Mount Sinai Fertility, Division of Reproductive Sciences, University of Toronto, Toronto, ON, Canada

OriginElle Fertility Clinic and Women's Health Centre, QC, Canada

S. L. Tan (✉)
OriginElle Fertility Clinic and Women's Health Centre, QC, Canada

MUHC Reproductive Centre, Department of Obstetrics and Gynecology, McGill University, QC, Canada
e-mail: sltan@originelle.com

J. G. Schenker et al. (eds.), *Hot Topics in Human Reproduction*, Reproductive Medicine for Clinicians 3, https://doi.org/10.1007/978-3-031-24903-7_16

popularization of IVM first emerged from work from McGill showing that a single injection of hCG with no other stimulation was able to achieve a live birth rate/cycle started of 40% [6]. Most recently, a consensus has emerged to promote more widespread expertise and training for IVM given its overall safety and reduced patient burden compared to hyperstimulation protocols with IVF [3].

Oocyte Physiology

Oocyte maturation is the physiological event that precedes, and is required for, successful fertilization and embryo development. Oocytes initially mature during the foetal period and become arrested at the diplotene stage of prophase I (GV stage) until they are committed to ovulation or atresia. Resumption of meiosis and progression through maturation result in arrest at the metaphase II stage, with extrusion of the first polar body (Fig. 16.1). In vivo, the trigger for resumption of meiosis is the preovulatory surge of LH. However, removal of the oocyte from the inhibitory influence of the follicle also allows spontaneous resumption of meiosis. Importantly, the success of IVM relies on techniques that promote both nuclear and cytoplasmic maturation of the oocyte. Nuclear maturation consists of germinal vesicle breakdown (induced by the LH surge in vivo) followed by resumption of meiosis and extrusion of the first polar body (MII). Cytoplasmic maturation is more difficult to assess microscopically as it involves the redistribution of various organelles, including cortical granules, and accumulation of factors that prepare the oocyte for fertilization and embryonic development [7].

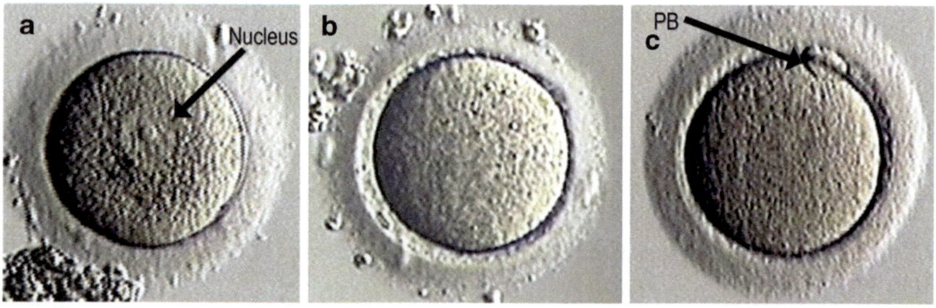

Fig. 16.1 Different stages of oocyte development; (**a**) Oocyte at germinal vesicle stage (GV), the nuclear membrane is intact and the nucleolus is visible; (**b**) Metaphase I oocyte, the nuclear membrane dissolves and the nucleolus disappears; (**c**) Metaphase II oocyte, the first polar body extrudes. Note: these oocytes are after the removal of cumulus cells

Potential Indications for IVM

IVM offers several key advantages over other assisted reproduction techniques, including a lower risk of adverse events and reduced financial and emotional burden owing to the short duration of monitoring and relatively low medication requirements. Historically, IVM has primarily been proposed as an alternative to IVF particularly for PCOS patients at risk for ovarian hyperstimulation syndrome (OHSS) given that they generally have the highest number of antral follicles for potential retrieval and simultaneously at highest risk of developing OHSS under traditional stimulation protocols. Other indications include patients with limited time for ovarian stimulation as well as those with contraindications to sustained elevation of oestradiol (E2). However, studies have shown that IVM can be used in almost all areas where IVF and other assisted reproductive techniques are used [2, 3, 8].

IVM in the Era of Antagonist Protocol with Agonist Trigger

Despite increasingly widespread use of GnRH agonist trigger in GnRH antagonist cycles, OHSS remains an ever-present risk with gonadotropin stimulation [9]. Indeed, early severe OHSS can occur if there is agonist trigger with low-dose hCG (1500 IU) rescue [9], if there is agonist trigger with high-dose oestrogen and progesterone supplementation for fresh ET [10, 11], or even if there is agonist trigger and a freeze all embryos policy. Only IVM can avoid OHSS completely. Another recent development of pseudo double-lumen needle (Steiner-Tan needle) allows even better results for IVM egg collection [12].

IVM Instead of Natural Cycle IVF

Classical natural cycle IVF involves no ovarian stimulation, and triggering is generally performed once the leading follicle reaches 18-mm diameter; however, it is associated with up to 30% risk of premature LH surge and ovulation. Alternatively, modified natural cycle IVF requires daily GnRH antagonist and FSH injections once a follicle reaches 14 mm and continued until hCG triggering, which equates to at least three FSH and GnRH antagonist injections with a subsequent 15–20% clinical pregnancy rate per cycle started given that only one MII oocyte is retrieved. In stimulated IVM, three injections of FSH are given on days 4, 6, and 8, with subsequent hCG administration when the leading follicle is 12–14-mm, thereby forgoing the need of GnRH antagonist. Furthermore, several MII and GV oocytes can be obtained to generate multiple blastocysts, thereby increasing clinical pregnancy rates per cycle started of up to 45–50% in women up to 37 years of age. In some cases, over 100 oocytes can be obtained at a single collection [13].

IVM Instead of FSH/IUI

The first line of treatment in many fertility programs is FSH stimulation combined with IUI. This treatment requires approximately eight to ten daily injections of FSH and has a pregnancy rate of 15–20% and multiple pregnancy rate of 30% and may cost upwards of $2000 in North America for medications and IUI. In comparison, IVM costs ~$4000 but requires fewer injections and generally yields higher pregnancy rates, lower odds of multiple pregnancy, and no risk of OHSS. In fact, Hatirnaz et al. have achieved a 35% live birth rate per IVM cycle with elective single ET. This group has even achieved good results using letrozole 5 mg a day for 5 days in early to mid-follicular phase and hCG trigger at 12–14 mm [14] and eSET [14].

In fact, the group from Perth has even shown that although fresh ET from IVM cycles are lower than from IVF cycle, if the embryos are frozen, the FERC pregnancy rates for both IVM and IVF are comparable [15].

IVM and Fertility Preservation for Cancer

When presented with an oncology patient wanting fertility preservation, it is important to first determine whether ovarian stimulation can be safely performed and how long the patient can wait before the start of chemotherapy [16]. If hormone stimulation is not contraindicated and chemotherapy can wait, ovarian stimulation followed by mature egg collection should be performed. However, IVM with or without ovarian tissue freezing are the only viable options if hormone stimulation is contraindicated or there is no time. Importantly, IVM allows multiple egg collections to be performed at any phase of the menstrual cycle including the luteal phase [17]. In the past few years, there has been a lot of work in this field extending the area of IVM and cancer fertility preservation. For example, it has been shown to allow fertility preservation for breast cancer [18], and there is even greater promise to combine in vitro growth and in vitro maturation [19]. Other advances in this area involve the addition of L-carnitine and B-glutathione to the IVM media to increase survival of vitrified GV oocytes and fertilization to become blastocysts, at least in the animal model [20, 21].

IVM and Resistant Ovary Syndrome

A few small studies and case reports have investigated the utility of IVM amongst women with repeated ART failure owing to resistant ovary syndrome with promising results [22, 23]. For instance, Galvao et al. reported a case series of 9 women with repeated IVF failure who underwent 24 IVM cycles and achieved a live birth rate of 16.7% per started cycle and 33.3% per patient [22].

Safety Concerns

Although the long-term developmental outcomes of children conceived with IVM have only been studied in small numbers, current evidence suggests that foetal outcomes and incidence of congenital malformations are similar to pregnancies derived from IVF and spontaneous pregnancies in healthy women [24–31]. This reflects related cellular and molecular studies which found normal ultrastructural morphology by transmission electron microscopy (TEM) and no increase in imprinting errors rates at maternally or paternally methylated gene loci in IVM-derived oocytes [32, 33].

Conclusion

IVM is a well-studied and safe procedure that has been practiced for several decades. It is a low intervention, mild approach to ART that offers improved safety and a simplified clinical approach compared to IVF [34, 35]. However, adoption has been mixed as incentives leading to enhanced uptake of IVM in the ART clinic vary widely around the world. As further innovations improve the recovery rate of immature oocytes and embryo yield from IVM, REI physicians may soon have an obligation to offer this technology more widely given its potential advantages to offer new approaches to infertility management and social fertility preservation and towards a new clinical paradigm of minimal or zero stimulation ART. When the senior author of this chapter started his career with Robert Edwards, Edwards emphasized that he had started his career in IVM and always held the dream that one day, women would have the same success rates with IVM as with IVF. He used to say, drugs only benefit big pharma. No woman would willingly give herself daily injections of hormones if she believed she could have the same success rates without having to use drugs [2].

Acknowledgements The authors acknowledge Dr. Mingju Cao for technical assistance in preparation of the manuscript.

References

1. Tan SL, Betts J, Mason B, Edwards RG, Tan SL, Campbell S, et al. Cumulative conception and live birth rates after in-vitro fertilisation. Lancet. 1992;339(8806):1390–4.
2. Niederberger C, Pellicer A, Cohen J, et al. Forty years of IVF. Fertil Steril. 2018;110(2):185–324.e5.
3. Hillson K, Clemens SC, Madhi SA, Voysey M, Pollard AJ, Minassian AM. In vitro maturation: a committee opinion. Fertil Steril. 2020;115(2):298–304. https://doi.org/10.1093/humrep/dew109.
4. Cha KY, Koo JJ, Ko JJ, Choi DH, Han SY, Yoon TK. Pregnancy after in vitro fertilization of human follicular oocytes collected from nonstimulated cycles, their culture in vitro and their transfer in a donor oocyte program**Prize paper, presented at the 45th Annual Meeting of The American Fertility Society, San Francisco, California, November 11 to 16, 1989. Fertil Steril. 1991;55(1):109–13.

5. Trounson A, Wood C, Kausche A. In vitro maturation and the fertilization and developmental competence of oocytes recovered from untreated polycystic ovarian patients*. Fertil Steril. 1994;62(2):353–62.
6. Chian RC, Gülekli B, Buckett WM, Tan SL. Priming with human chorionic gonadotropin before retrieval of immature oocytes in women with infertility due to the polycystic ovary syndrome. New Engl J Med. 1999;341(21):1624–6. https://doi.org/10.1056/NEJM199911183412118.
7. Chian RC, Buckett WM, Tan SL. In-vitro maturation of human oocytes. Reprod Biomed Online. 2004;8(2):148–66.
8. Plancha CE, Rodrigues P, Marques M, Almeida JM, Navarro-Costa P. The time is ripe for oocyte in vitro maturation. J Assist Reprod Gen. 2021;38(6):1281–3.
9. Seyhan A, Ata B, Polat M, Son WY, Yarali H, Dahan MH. Severe early ovarian hyperstimulation syndrome following GnRH agonist trigger with the addition of 1500 IU hCG. Hum Reprod. 2013;28(9):2522–8. https://doi.org/10.1093/humrep/det124.
10. Fatemi HM, Popovic-Todorovic B, Humaidan P, Kol S, Banker M, Devroey P, et al. Severe ovarian hyperstimulation syndrome after gonadotropin-releasing hormone (GnRH) agonist trigger and "freeze-all" approach in GnRH antagonist protocol. Fertil Steril. 2014;101(4):1008–11. https://doi.org/10.1016/j.fertnstert.2014.01.019.
11. Gurbuz AS, Gode F, Ozcimen N, Isik AZ. Gonadotrophin-releasing hormone agonist trigger and freeze-all strategy does not prevent severe ovarian hyperstimulation syndrome: a report of three cases. Reprod Biomed Online. 2014;29(5):541–4. https://doi.org/10.1016/j.rbmo.2014.07.022.
12. Rose BI, Laky D. A comparison of the Cook single lumen immature ovum IVM needle to the Steiner-Tan pseudo double lumen flushing needle for oocyte retrieval for IVM. J Assist Reprod Genet. 2013;30(6):855–60. https://doi.org/10.1007/s10815-013-0006-1.
13. Dahan MH, Ata B, Rosenberg R, Chunh JT, Son WY, Tan SL. Collection of 125 oocytes in an in vitro maturation cycle using a new oocyte collection technique: a case report. J Obstet Gynaecol Can. 2014;36(10):900–3. http://www.jogc.com/article/S1701216315304394/abstract
14. Hatırnaz S, Hatırnaz E, Dahan MH, Tan SL, Ozer A, Kanat-Pektas M, et al. Is elective single-embryo transfer a viable treatment policy in in vitro maturation cycles? Fertil Steril. 2016;106(7):1691–5.
15. Walls ML, Hunter T, Ryan JP, Keelan JA, Nathan E, Hart RJ. In vitro maturation as an alternative to standard in vitro fertilization for patients diagnosed with polycystic ovaries: a comparative analysis of fresh, frozen and cumulative cycle outcomes. Hum Reprod. 2015;30(1):88–96. https://doi.org/10.1093/humrep/deu248.
16. Chian RC, Huang JY, Gilbert L, Son WY, Holzer H, Cui SJ, et al. Obstetric outcomes following vitrification of in vitro and in vivo matured oocytes. Fertil Steril. 2009;91(6):2391–8. https://doi.org/10.1016/j.fertnstert.2008.04.014.
17. Demirtas E, Elizur SE, Holzer H, Gidoni Y, Son WY, Chian RC, et al. Immature oocyte retrieval in the luteal phase to preserve fertility in cancer patients. Reprod Biomed Online. 2008;17(4):520–3.
18. Shalom-Paz E, Almog B, Shehata F, Huang J, Holzer H, Chian RC, et al. Fertility preservation for breast-cancer patients using IVM followed by oocyte or embryo vitrification. Reprod Biomed Online. 2010;21(4):566–71. https://doi.org/10.1016/j.rbmo.2010.05.003.
19. Telfer EE, Andersen CY. In vitro growth and maturation of primordial follicles and immature oocytes. Fertil Steril. 2021;115(5):1116–25.
20. Moawad AR, Tan SL, Xu B, Chen HY, Taketo T. L-carnitine supplementation during vitrification of mouse oocytes at the germinal vesicle stage improves preimplantation development following maturation and fertilization in vitro. Biol Reprod. 2013;88(4):104. https://doi.org/10.1095/biolreprod.112.107433.
21. Moawad AR, Xu B, Tan SL, Taketo T. l-carnitine supplementation during vitrification of mouse germinal vesicle stage-oocytes and their subsequent in vitro maturation improves meiotic spindle configuration and mitochondrial distribution in metaphase II oocytes. Hum Reprod. 2014;29(10):2256–68. https://doi.org/10.1093/humrep/deu201.

22. Galvão A, Segers I, Smitz J, Tournaye H, Vos MD. In vitro maturation (IVM) of oocytes in patients with resistant ovary syndrome and in patients with repeated deficient oocyte maturation. J Assist Reprod Gen. 2018;35(12):2161–71.
23. Kornilov NV, Pavlova MN, Yakovlev PP. The live birth in a woman with resistant ovary syndrome after in vitro oocyte maturation and preimplantation genetic testing for aneuploidy. J Assist Reprod Gen. 2021;38(6):1303–9.
24. Ho VNA, Braam SC, Pham TD, Mol BW, Vuong LN. The effectiveness and safety of in vitro maturation of oocytes versus in vitro fertilization in women with a high antral follicle count. Hum Reprod. 2019;34(6):1055–64.
25. Cha KY, Chung HM, Lee DR, Kwon H, Chung MK, Park LS, et al. Obstetric outcome of patients with polycystic ovary syndrome treated by in vitro maturation and in vitro fertilization–embryo transfer. Fertil Steril. 2005;83(5):1461–5.
26. Fadini R, Renzini MM, Guarnieri T, Canto MD, Ponti ED, Sutcliffe A, et al. Comparison of the obstetric and perinatal outcomes of children conceived from in vitro or in vivo matured oocytes in in vitro maturation treatments with births from conventional ICSI cycles. Hum Reprod. 2012;27(12):3601–8.
27. Yu EJ, Yoon TK, Lee WS, Park EA, Heo JY, Ko YK, et al. Obstetrical, neonatal, and long-term outcomes of children conceived from in vitro matured oocytes. Fertil Steril. 2019;112(4):691–9.
28. Mostinckx L, Segers I, Belva F, Buyl R, Santos-Ribeiro S, Blockeel C, et al. Obstetric and neonatal outcome of ART in patients with polycystic ovary syndrome: IVM of oocytes versus controlled ovarian stimulation. Hum Reprod. 2019;34(8):1595–607.
29. Roesner S, von Wolff M, Elsaesser M, Roesner K, Reuner G, Pietz J, et al. Two-year development of children conceived by IVM: a prospective controlled single-blinded study. Hum Reprod. 2017;32(6):1341–50.
30. Buckett WM, Chian RC, Holzer H, Dean N, Usher R, Tan SL. Obstetric outcomes and congenital abnormalities after in vitro maturation, in vitro fertilization, and intracytoplasmic sperm injection. Obstet Gynecol. 2007;110(4):885–91.
31. Söderström-Anttila V, Salokorpi T, Pihlaja M, Serenius-Sirve S, Suikkari AM. Obstetric and perinatal outcome and preliminary results of development of children born after in vitro maturation of oocytes. Hum Reprod. 2006;21(6):1508–13.
32. Kuhtz J, Romero S, Vos MD, Smitz J, Haaf T, Anckaert E. Human in vitro oocyte maturation is not associated with increased imprinting error rates at LIT1, SNRPN, PEG3 and GTL2. Hum Reprod. 2014;29(9):1995–2005.
33. Coticchio G, Canto MD, Fadini R, Renzini MM, Guglielmo MC, Miglietta S, et al. Ultrastructure of human oocytes after in vitro maturation. Mhr Basic Sci Reprod Med. 2016;22(2):110–8.
34. Teede HJ, Misso ML, Costello MF, Dokras A, Laven J, Moran L, et al. Recommendations from the international evidence-based guideline for the assessment and management of polycystic ovary syndrome†‡. Hum Reprod Oxf Engl. 2018;33(9):1602–18.
35. Costello MF, Misso ML, Balen A, Boyle J, Devoto L, Garad RM, et al. Evidence summaries and recommendations from the international evidence-based guideline for the assessment and management of polycystic ovary syndrome: assessment and treatment of infertility. Hum Reprod Open. 2019;2019(1):hoy021.

Medical and Ethical Aspects of Noninvasive Prenatal Diagnosis (NIPT)

17

Wolfgang Holzgreve

Introduction

As part of the science and technology revolution especially from the middle of the mid-nineteenth century, medicine experienced an enormous progress with, e.g., an increase of life expectancy from an average of less than 50 years in the year 1950 to more than 70 years in 2020. For many people, the quality of life has also been improved, including for those with congenital and chronical diseases [1]. Although there is still an enormous discrepancy in living conditions between countries and regions, the result of this progress benefits many people worldwide today. At the same time, medical advances have made it possible for many people to live a more self-determined life than ever before, especially multimorbid ill and persons with handicap. But there is hardly a success story without new risks which have to be watched generally and in perinatal medicine in particular, because during pregnancy two persons are involved, mother and child.

Since we have historical recordings of human thoughts and emotions, there is evidence for concern of expectant parents regarding the health of their unborn children. In the past, however, the ability to find out whether the growing fetus had problems in utero was very limited. Especially in the 1970s, this changed with the introduction of diagnostic ultrasound which allowed to visualize the unborn child without harm, and around the same time, biochemical marker screening approaches were developed first for the prediction of neural tube defects and later for chromosomal anomalies in the fetus and embryo. The new powerful screening tests have to be combined in a logical and affordable way [2], and their proper use now requires well-informed up-to-date counseling.

W. Holzgreve (✉)
University Medical Center, Bonn, Germany
e-mail: wolfgang@holzgreve.net

J. G. Schenker et al. (eds.), *Hot Topics in Human Reproduction*, Reproductive Medicine for Clinicians 3, https://doi.org/10.1007/978-3-031-24903-7_17

History of Prenatal Diagnosis

Outstanding and significant improvements of the ultrasound technology quickly allowed amniocentesis to be performed around 16 weeks of pregnancy and since the middle of the 1980s chorionic villus biopsy already at around 10 weeks of gestation [3]. Ultrasound, however, will most likely never be replaced in pregnancy surveillance by genetic techniques, because the majority of the structural defects to be detected prenatally, which constitute the majority of the 2–4% congenital anomalies, are multifactorial (e.g., cardiac or neural tube defects).

Cultures from amniotic fluid or chorionic villus cells enabled cytogenetic analysis, and prenatal diagnosis of trisomy 21 was offered to women with an age of 35 years and older because of their increased risk for fetal aneuploidies so regularly that unfortunately and wrongly this offer was perceived as almost synonymous with "prenatal diagnosis" in general. From the beginning, however, the ability to have a legal termination of pregnancy based on a prenatal diagnosis raised the serious conflict between the fundamental prohibition, based on the concept of dignity of every (born and unborn) human being, of actively ending any human life on the one side and the autonomy of the pregnant woman on the other. This basic conflict ultimately cannot be solved in a absolutely satisfactory way, and different countries found various solutions to deal with the dilemma.

If we look into the fascinating history of prenatal diagnosis in more detail, we recognize remarkable progress in prenatal screening and diagnosis within a relatively short time, e.g., that after maternal age was for a long time the only parameter used to define an increased risk for a chromosomal anomaly in a fetus, the observation of lower maternal serum alpha-fetoprotein (AFP) levels being associated with trisomy 21 in populations screened for neural tube defects initiated a systematic search for other biochemical markers in the maternal blood suitable to assess the risk for common trisomies. Using AFP, free β-human chorion gonadotropin (hCG) and unconjugated estriol ("triple test"; sometimes complemented by other markers) in the second trimester of pregnancy multi-parameter maternal serum testing became the recommended and widely used approach to assess an individual risk for fetal trisomies. A further significant increase in sensitivity was achieved by including the extension of the skin in the fetal neck (nuchal translucency) as determined by ultrasound in the first trimester of pregnancy and the simultaneous measurement of hCG and the pregnancy associated plasma protein A (PAPP-A), known as the first trimester or combined test. Various additional serum and ultrasound markers as well as testing approaches were proposed [4], but the major step forward in screening was the development of NIPT which is now by far the best screening method for fetal trisomy together with applying ultrasound apart from testing for trisomies. NIPT can also be applied new for structural chromosomal anomalies (submicroscopical chromosomal aberrations and microdeletions) which are not age-dependent.

Another burden on prenatal diagnosis by amniocentesis or CVS always was the low, but definitely existing risk of harming the pregnancy by the prenatal invasive sampling procedure [5]. Therefore since the 1970s, there was an intense search for a noninvasive method of prenatal diagnosis, and in a collective effort of international research first the isolation of fetal cells from the blood of pregnant women was tried, whereas later this method became successful and clinically mature by looking at cell-free DNA in the maternal blood. Because progress in medicine through publications, lectures, and media travels fast these days, now more than ten million cases have been investigated noninvasively by the NIPT methodology.

The introduction of NIPT is a positive example for how a new technology should be introduced into clinical practice, because after the research and development after careful planning [6], the proper trials were performed [7–9], so that the public was prevented from an immature technical approach spreading into clinical use without rigorous evaluation.

Techniques of NIPT

Further to the now well-developed techniques to test for trisomies 21, 18, and 13 prenatally by NIPD, prenatal diagnosis of trisomies [10, 11] is now possible by a variety of techniques such as RTQ-PCR, digital PCR, and more recently HTS and some of these approaches are applicable additionally for a number indications such as paternally inherited dominant conditions or autosomal recessive conditions with different mutations in both parents and new mutations [12]. Testing, for example, of FGFR3 (achondroplasia, thanatophoric dysplasia) or FGFR2 (Apert syndrome) is a noninvasive diagnostic option, e.g., in cases with suspicious ultrasound findings when the parents are phenotypically normal [13]. The diagnostic approach of course is more challenging in recessive conditions when both parents carry the same mutation as well as for maternal dominant disorders or X-linked conditions when the mother is a carrier. Different approaches have been applied to address these situations, e.g., by the relative mutation dosage as assessed by droplet digital PCR (ddPCR) using slight differences in the ratio of mutant and wild-type alleles in the cfDNA depending on the presence of absence of mutant alleles in the cffDNA. The second strategy is HTS-based genome-wide SNP-genotyping, and the assessment of relative haplotype dosage in theory permitting the testing of virtually any monogenic condition [14]. This indirect testing approach uses SNP haplotypes linked to a specific locus of interest and is reliable due to the abundancy of available SNPs. For prenatal diagnosis, the paternal haplotype in the cfDNA is detected by the analysis of SNPs homozygous in the mother (AA) and heterozygous in the father (AC). The statistical significance of any allelic imbalance is calculated by a sequential probability ratio test (SPRT). Obviously, the parental coupling phase (i.e., the haplotype linked to the mutation) has to be known, e.g., by testing family members in one or both. The genome-wide approach can be customized by recent targeted capture sequencing technologies to restrict sequencing to genomic regions of interest [15]. An alternative approach was proposed by Dan et al. [16] by searching for known or de novo variants following HTS-based noninvasive targeted capture sequencing of cfDNA and of both parents.

Different studies have shown, however, that both professionals and pregnant women find it difficult to grasp the concept of an expanded NIPT. In the meantime, the offer of an expanded NIPT has already started in many developed countries, including the United States besides Belgium and the Netherlands [17]. More care providers offer their patients commercially available expanded NIPT including besides the three trisomies also submicroscopical chromosomal aberrations and microdeletions.

Ethical Considerations in Prenatal Medicine

A major problem of prenatal medicine from the beginning was the fact that in the majority of cases the prenatally diagnosed chromosomal and metabolic diseases could not be treated. Therefore Sir William Liley, one of the pioneers in this field, even spoke about a "search and destroy" mission. He was, however, at the same time "the father" of prenatal therapy, because following the ability to assess by amniocentesis and blood sampling in utero the severity of a rhesus blood group incompatibility, he developed techniques for treating the affected babies prenatally by intrauterine blood transfusion with the delicate technique of fetoscopy, therefore obtaining just a precise diagnosis mutated into effective though risky prenatal therapy and followed this way in the general principle in medicine that only diagnosis should not be considered as the ultimate goal. In order to highlight this positive development already in 1987, I gave a book about these newly developing techniques the title *Prenatal Medicine* rather than just "Prenatal Diagnosis" [18]. Therefore the progress in prenatal medicine, which for a long time was moving way ahead of the possibilities in prenatal therapy, now is bridging the gap with significant achievements in prenatal therapy such as intrauterine surgery or gene therapy [19].

Although prenatal diagnosis was looked at skeptically because of ethical concerns in some parts of the general population from the beginning, it was soon confirmed that it has saved many lives of unborn children, because less often than before by these technological advances, women with anxieties could be calmed based on the proper prenatal diagnosis, e.g., in cases of an in utero rubella virus infection or when women became pregnant with so-called advanced maternal age. Regarding genetic diseases with Mendelian patterns of inheritance, Bernadette Modell in the United Kingdom showed that in populations with increased risk for beta-thalassemia after the proper counseling about the 25% recurrence risk, the pregnancy rates went down dramatically, whereas they increased to normal again after the possibility of prenatal diagnosis even by the invasive method of fetoscopic blood sampling [5].

Noninvasive prenatal testing (NIPT) is not only a major breakthrough and progress for the pregnant women concerned but even helps to understand better pregnancy-related diseases such as preeclampsia and autoimmune diseases [6]. Not surprisingly, however, the general discussion about the justification for prenatal diagnosis of untreatable conditions started immediately again when this progress on noninvasive testing became available, sometimes mixed up with the progress of the noninvasive over the invasive techniques.

For example, the new option of gene therapy has to be taken into account in those genetic conditions of the fetus, which now can be treated much better such as hemophilia by factor substitution, innovative medicines, or gene therapy. This example of progress in gene therapy is therefore also covered by Johannes Oldenburg [20].

In an assessment of genome-wide screening, the WHO has concluded that, although research should always be encouraged, the benefits vs harms of implementation of GW cfDNA screening must be weighed carefully. Healthcare providers and grant-awarding bodies have a responsibility to ensure that more robust data and management strategies are available before endorsing studies or strategies incorporating GW-cfDNA testing into nationally reimbursed screening programs.

Together with Belgium, the Netherlands is the only other country in Europe where NIPT is offered as first-tier screening test and in the genome-wide option [21]. The background for this is in the Netherlands (NL) a unique law, the Population Screening Act, which regulates screening for untreatable diseases [22]. The law aims at protecting citizens against the potential negative effects of screening. Each time a new screening for these conditions is proposed, a governmental license is required before the screening can be implemented [23].

This explains why implementation of prenatal screening (PS) in the Netherlands has always been careful and thoughtfully weighed, with special attention for its potential side effects. The foundation on which prenatal screening has been endorsed in the NL is that the offer should enhance reproductive choices. The national screening program in the Netherlands was implemented in 2007 and from the outset all pregnant women counseled regarding the following screening options: the combined testing (CT) as screening for trisomy 21 (Down syndrome) and a structural ultrasound examination at around 20 weeks as screening for structural anomalies. In 2011, screening for trisomies was extended to trisomies 13 and 18. Whereas mid-trimester US screening is free, women have to pay 165 Euro's for aneuploidy screening. Prof. Catarina M. Bilardo (personal communication) concluded that this fact may have contributed to the low uptake of screening for trisomies.

Obviously regarding medical benefit and safety, NIPT is entirely advantageous over other available prenatal diagnostic tests such as amniocentesis or chorionic villus biopsy: While these invasive interventions involve a small risk of procedure-induced miscarriage, NIPT requires nothing but a blood sample from the pregnant woman and therefore poses no risk to her pregnancy. Due to its noninvasive nature, it is physically and psychologically much less burdensome, and it can be conducted earlier in pregnancy and more discretely. It is, however, just NIPT's procedural harmlessness or triviality that in the eyes of critics gives rise to some specific ethical concerns of which the fears of "routinization" and "pressure" seem the most prominent rhetoric coatings, as Bettina Schöne-Seifert and Chiara Junker defined it well [24]: "That mindful decision-making should be a key educational goal (not only) of NIPT counseling which could be achieved through stepwise disclosure and indirect social pressure could be the most plausible threat to reproductive free choice. While continuous efforts need to be made to prevent such pressure—not least by ensuring balanced availability of options—restricting testing options, and thus freedom of choice, cannot be the answer to this concern. Lastly, we suggest abandoning the vague term 'routinization' and instead focusing on specified concerns to enable a fruitful debate". Objections to the provision and the use of NIPT are usually not directed against the test itself, but against its role as a promoter of selective abortions. Notwithstanding some uneasiness with the term 'selection' especially in German context with its history of euthanasia during the Nazi times.

The ethicist Christoph Rehmann-Sutter stated [25]: "Bioethics needs to deal with all levels of social practices of technology use: the regulatory and the individual, the social and the intergenerational, its place in national history and the comparison with other traditions. It needs to address and acknowledge all perspectives involved: the professionalism of physicians and experts involved in reproductive genetics, the concerns of women and their partners, the perspectives of families and their generations, and of course the regulators' arguments that raise controversy in different cultural contexts. Such an ethics creates a respectful space of mutual understandings and public deliberation that helps to make 'personalized ethics' [26] a less solitary enterprise."

In 2001, long before noninvasive prenatal testing (NIPT) was available, the German Parliament discussed about "Law and Ethics of Modern Medicine and Biotechnology" and at that time the German Member of Parliament, Andrea Fischer stated, about the ethical implications of NIPT: "None of us should arrogantly place our own morals above others. Each of us should let ourselves be unsettled by the arguments of the other in this discussion" [27].

The German Ethics Council (Deutscher Ethikrat) stated in 2013: "The majority of the members are of the opinion that a non-invasive prenatal genetic diagnosis … should only be carried out if there is an increased risk of a genetic disease or malformation." Whereas the Federal Joint Committee in Germany concluded: "…that NIPT can be used at the expense of the health insurance if the question arises as part of the medical care for pregnant women whether a fetal trisomy could be present and this represents an unreasonable burden for the pregnant woman" [27].

Offering NIPT as a first-tier screening test may lead to routinization. This concept refers to the fact that women may embark on prenatal screening more superficially as there is less risk of having to undergo an invasive procedure implying a (small) risk of miscarriage [28]. This may also lead to societal pressures to participate in prenatal screening and to stigmatization of women, families who do not do it [29]. Moreover, it has also been suggested that it may be easier to decide for termination of pregnancy of an affected fetus at an earlier stage in pregnancy, as compared to after an amniocentesis performed at 16 weeks [30]. However, all these concerns still lack empirical evidence and may not be founded.

Another argument says that persons with disabilities might face increased discrimination in their lives as a result of NIPT. While such discrimination could in principle take the form of intentionally hostile behavior, there is even a stronger concern about indirect discrimination. In particular that it might become harder for persons with disabilities to find specialized health care experts. If fewer patients with particular needs exist in the future, this might lead to fewer medical experts in that field and subsequently to lower-quality healthcare. This would be a form of indirect discrimination: no one would willingly try to worsen the quality of healthcare for persons with trisomy 21, but it would be an indirect effect of the lower demand [31, 32].

The most common response to this worry says that a termination of pregnancy is not discriminatory if it occurs on the ground that the parents want to make use of their reproductive choices or because the mother believes that she would not be able to take care of the special needs of a child with a disability.

The most commonly voiced concern regarding the parents of children with disabilities is that they might hear remarks blaming them for having a child with a disability, that they might be faced with attitudes that the social responsibility for their children should primarily lie with them, or that their children should live more or less separately from the rest of society. There is a consensus in the medical ethics literature that such reactions and attitudes are inappropriate. Even though this may still occur occasionally, it is often argued that there is no empirical evidence that there is now more discrimination of this kind than before the introduction of NIPT. Quite to the contrary, there is reason to believe that the social acceptance of children with disabilities has actually increased [33].

For a majority in many populations, a way to deal with this dilemma about free choice of women for prenatal tests and the avoidance to go on the path to even termination of pregnancy is to institute knowledgeable and empathetic counseling about the facts of prenatal diagnosis and its consequences so that women can make their own informed choices within the legal limits of their society. This counseling, however, not only has to be empathetic and understandable but also has to take the medial progress into account, e.g., in utero or postpartum therapeutic options in congenital diseases. For example, the new option of gene therapy has to be taken into account in those genetic conditions of the fetus, which now can be treated much better such as treatment of SMA by innovative medicines or gene therapy [34].

Overall, since the number of invasive procedures has greatly been reduced by NIPT, this technique has already worldwide saved many fetal losses caused by invasive procedures, although there risk of amniocentesis and CVS is probably low now in experienced hands. The fascinating development of NIPT is an example that significant progress in the laboratory after rigorous testing can enter clinical routine. NIPT has improved the choices significantly for women and couples in a sensitive area of medicine—but we always have to proceed with caution making use of true progress but at the same time preventing negative side effects.

References

1. United Nations Development Programme. Human Development Reports. Development of the Human Development Index (HDI) from 1990 to 2013. http://hdr.undp.org/en/indicators/137506. Accessed 11 Jan 2021
2. Holzgreve W. Will ultrasound-screening and ultrasound-guided procedures be replaced by non-invasive techniques for the diagnosis of fetal chromosome anomalies? Ultrasound Obstet Gynecol. 1997;9:217–9.

3. Holzgreve W, Miny P. Chorionic villus sampling and placental biopsy. In: James DK, Steer PJ, Weiner CP, Gonik B, editors. High risk pregnancy. 2nd ed. Philadelphia: Saunders; 1999. p. 207–14.
4. Holzgreve W. Prenatal screening and diagnosis. In: Fetal medicine: basic science and clinical practice. Edited by C. Rodeck und M. Whittle. Edinburgh: Churchill Livingstone; 2006. p. 1162.
5. Holzgreve W, Tercanli S, Surbek D, Miny P. Invasive diagnostic methods. In: Rodeck C, Whittle MJ, editors. Fetal medicine: basic science and clinical practice. 2nd ed. Churchill Livingstone Publishers; 2006. p. 417–33.
6. Holzgreve W, Garritsen HSP, Tercanli S, Miny P, Nippert I, Gänshirt-Ahlert D. Noninvasive prenatal diagnosis. Strategy for a clinical trial. In: Simpson JL, Elias S, editors. Annals of the New York Academy of Sciences, vol. 731. New York: The New York Academy of Sciences; 1994. p. 253–6.
7. Holzgreve W, Hahn S. Fetal DIagnosis. In: Cooper DN, editor. Encyclopedia of the human genome, vol. 2. Nature Publishing Group; 2003. p. 477–80.
8. Bianchi DW, Simpson JL, Jackson LG, Elias S, Holzgreve W, Evans MI, Dukes KA, Sullivan LM, Klinger KW, Bischoff FZ, Hahn S, Johnson KL, Lewis D, Wapner RJ, de la Cruz F. Fetal gender and aneuploidy detection using fetal cells in maternal blood: analysis of NIFTY I data. Prenat Diagn. 2002;22:609–15.
9. Bischoff FZ, Hahn S, Johnson KL, Simpson JL, Bianchi DW, Lewis DE, Weber WD, Klinger K, Elias S, Jackson LG, Evans MI, Holzgreve W, de la Cruz F. Intact fetal cells in maternal plasma: are they really there? Lancet. 2003;361:139–40.
10. Lo YM, Corbetta N, Chamberlain PF, Rai V, Sargent IL, Redman CW, Wainscoat JS. Presence of fetal DNA in maternal plasma and serum. Lancet. 1997;1997(350):485–7.
11. Zimmermann B, Levett L, Holzgreve W. Use of real-time polymerase chain reaction for the detection of fetal aneuploidies. Methods Mol Biol. 2006;336:83–100.
12. Zhong XY, Hahn S, Holzgreve W. Prenatal identification of fetal genetic traits. Lancet. 2001;357:310–1.
13. Chitty LS, Mason S, Barrett AN, Lench N, Daley R, Jenkins LA. Non-invasive prenatal diagnosis of achondroplasia and thanatophoric dysplasia, next-generation sequencing allows for a safer more accurate and comprehensive approach. Prenat Diagn. 2015;35:656–62.
14. Lo YM, Chan KCA, Sun H, Chen EZ, Jiang P, Lun FMF, Zheng YW, Leung TY, Lau RK, Cantor CR, Chiu RWK. Maternal plasma DANN sequencing reveals the genome—wide genetic and mutational profile of the fetus. Sci Transl Med. 2010;2, Glra 91
15. Hudecova I, Chief RWK. Non-invasive prenatal diagnosis of thalassemias using maternal plasma cell free DNA. Best Pract Res Clin Obstet Gynaecol. 2017;39:63–73.
16. Dan S, Yuan Y, Wang Y, et al. Non-invasive prenatal diagnosis of lethal skeletal dysplasia by targeted capture sequencing of maternal plasma. PLoS One. 2016;11:e0159355.
17. Bilardo CM. The implementation of NIPT in The Netherlands. J Perinat Med. 2021;49(8):941–4.
18. Holzgreve W. Pränatale Medizin. Heidelberg/New York: Springer Verlag; 1987. p. 186.
19. Holzgreve W. Fetal anomalies—from prenatal diagnosis to therapy. J Perinat Med. 2018;46(9):951–2. https://doi.org/10.1515/jpm-2018-0333.
20. Oldenburg J. Gene therapies of hemophilia—a difficult way and first success, Symposium on NIPT, Bonn; 2020.
21. Van der Meij KRM, Sistermans EA, Macville MVE, et al. TRIDENT-2: national implementation of genome-wide non-invasive prenatal testing as a first-tier screening test in the Netherlands. Am J Hum Genet. 2019;105(6):1091–101.
22. Population Screening Act. Wet bevolkingsonderzoek. The Hague; 1992.
23. Kater-Kuipers A, Bunnik EM, de Beaufort ID, Galjaard RJH. Limits to the scope of non-invasive prenatal testing (NIPT): an analysis of the international ethical framework for prenatal screening and an interview study with Dutch professionals. BMC Pregnancy Childbirth. 2018;18:409.
24. Schöne-Seifert B, Junker C. Making use of NIPT: rethinking issues of routinization and pressure. J Perinat Med. 2021;49(8):959–64.

25. Rehmann-Sutter C. Should prenatal screening be seen as 'selective reproduction'? Four reasons to reframe the ethical debate. J Perinat Med. 2021;49(8):953–8.
26. Meskus M. Personalized ethics: the emergence and the effects in prenatal testing. BioSocieties. 2012;7:373–92.
27. Huster S. Non-invasive prenatal diagnostics (NIPD) in the system of medical care. Ethical and legal issues. J Perinat Med. 2021;49(8):972–8.
28. RIVM. Combinatietest en NIPT. 2019. https://www.rivm.nl/down-edwards-patau-en-seo/combinatietest-en-nipt.
29. Di Renzo GC, Bartha JL, Bilardo CM. Expanding the indications for cell-free DNA in the maternal circulation: clinical considerations and implications. Am J Obstet Gynecol. 2019;220:537–42.
30. Dondorp W, de Wert G, Bombard Y, Bianchi DW, Bergmann C, Borry P, Chitty LS, Fellmann F, Forzano F, Hall A, Henneman L, Howard HC, Lucassen A, Ormond K, Peterlin B, Radojkovic D, Rogowski W, Soller M, Tibben A, Tranebjærg L, van El CG, Cornel MC, European Society of Human Genetics, American Society of Human Genetics. Non-invasive prenatal testing for aneuploidy and beyond: challenges of responsible innovation in prenatal screening. Eur J Hum Genet. 2015;23:1438–50.
31. Dufner A. Non-invasive prenatal testing (NIPT): does the practice discriminate against persons with disabilities. J Perinat Med. 2021;49(8):945–8.
32. Heinrichs B. Moral ambivalence. A comment on non-invasive prenatal testing from an ethical perspective. J Perinat Med. 2021;49(8):949–52.
33. German Ethics Council. The future of genetic diagnosis—from research to clinical practice. Opinion. Berlin: German Ethics Council; 2013.
34. Kirschner J. Postnatal gene therapy for neuromuscular diseases—opportunities and limitations. J Perinat Med. 2021;49(8):1011–5. https://doi.org/10.1515/jpm-2021-0435.

Embryo-Specific Communication and Interaction with Maternal Environment: Role of Preimplantation Factor (PIF*)

18

Eytan R. Barnea

Maternal Awareness of Embryo Presence

The 5–7 days delay post-fertilization until implantation is critical for reproduction as in IVF, a similar 4–5 days elapse post-transfer is observed [1]. Thus, in maternal recognition, preimplantation is silent clinically. Post-implantation as embryogenesis initiates will need significant maternal adaptation until term unless pathology occurs. Such adaptation requires major (metabolic, hormonal) changes and access to nutritional resources to sustain until delivery. We advance the concept that embryo is the driver of gestation and the maternal system a responsive, receptive entity. Post-implantation, the host plays a critical role to accept/reject and sustain/eliminate the embryo sooner or later up to full-term delivery [2]. The uterus is critical for favorable pregnancy outcome. However, extrauterine implantation is possible where also specific changes related to maternal recognition of pregnancy occur. Viewing the embryo as a "parasite," it will seek/find the environment where it can thrive. What makes the embryo so successful? Its messaging and survival skills are preserved even in the harshest maternal environment of pregnancy without a uterus.

Embryo Signaling: To Be Specific or Not to Be Specific

Pregnancy is a unique phenomenon not to be replicated by any other condition. The search for embryo-specific compounds that would initiate such determining embryo-maternal cross talk has been an ongoing quest. We aimed to identify embryo viability marker expressed throughout mammalian reproduction, which is not detected in the unfertilized egg, detected already in the two-cell stage embryo, present and functional in different mammalian species, culture media, and maternal serum, and

E. R. Barnea (✉)
Founder BioIncept LLC, New York, NY, USA

J. G. Schenker et al. (eds.), *Hot Topics in Human Reproduction*, Reproductive Medicine for Clinicians 3, https://doi.org/10.1007/978-3-031-24903-7_18

not found in non-pregnant mammals. Once identified, we named it preimplantation factor (PIF). Isolation was aided by LPBA bioassay using PIF, a multistep chromatographic isolation step of embryo culture media identifying short 15 AA peptide, while also a shorter 9 AA is also secreted [3].

PIF-Specific Marker of Embryo Viability

Embryo post-fertilization acts in a stealth mode, surrounded by the semi-permeable zona pellucida where maternal immune cells cannot access but secretory products from the embryo can be released successfully. As such, communication can initiate. The embryo/graft uses small footprint, while it develops and prepares for implantation/transplant acceptance [4]. PIF secreted by viable but not atretic embryos, two cells stage in mice, four cells human, and six cells bovine embryo cultures. Levels increase with progress to blastocysts [5, 6].

PIF-based embryo-maternal signaling is seamless due to its expression in the earliest trophoblast, at the implantation site (the extra-villous trophoblast—EVT) promoting effective embryo-maternal communication. PIF expression further increases in the first trimester trophoblast, and similarly maternal circulation where PIF levels do increase the major source is the placenta less from embryo [2]. Placental and in maternal circulation PIF levels plateaus in the second trimester declining afterwards.

Compared to non-pregnancy-specific hCG found in circulation up to 4 weeks after pregnancy PIF being a peptide that has short half-life, minutes [1]. PIF signaling enables embryo maternal communication via the placenta/uterus but also with the systemic maternal immunity. Placental PIF expression declines prematurely in IUGR, and prematurity and preeclampsia-reflect intimate PIF involvement with pregnancy physiology [5].

PIF Supports Embryo Development and Its Survival: Autocrine Effect

PIF beyond being secreted by viable embryos and reaching the maternal circulation also exert autocrine trophic and protective effects. This is logical since the early embryo is segregated in the fallopian tube. Therefore, the embryo develops by relying on endogenous compounds, so by the time it reaches the uterus, separation to embryoblast and trophoblast, "the leading edge," blastocyst takes place and is ready to implant.

We tested whether PIF can promote embryo development. Therefore sPIF (synthetic PIF) was developed obviating the native PIF use. In bovine IVF, embryos are cultured in large groups. To improve survival, PIF added to singly cultured embryos for first 3 days and after media changed embryos were cultured for additional 4 days without the peptide [6]. PIF promoted development of embryos up to the blastocyst stage. On the other hand, when murine embryos were cultured under optimal conditions, the addition of PIF did not further enhance their development. This strongly indicates that PIF action is adapted to need. This established the trophic effect of the peptide on the embryo. The next step was to determine whether, oppositely, minimizing PIF access to the embryo would impair development [7].

Recognizing that the embryo must fend for itself, the effect of PIF in the presence of adverse maternal environment was tested using serum of women with a history of recurrent pregnancy loss (RPL). Testing the effect of toxic serum even at 5% when added to the culture media led to significant embryo demise. In higher concentration of the serum, the demise was total. The results revealed that embryos can be affected in two different ways: delaying embryo development or causing their demise [8]. The former was due to presence of low <3 kDa toxins, while the exposure to the higher MW toxins was associated with embryo demise. PIF addition to the culture media negates both adverse pathologies. The data demonstrated that irrespective of whether the serum delayed embryo development or led to their demise, PIF acted as a rescue factor [8]. It begged the question: How does this take place? The studies using PIF in fluorescent form demonstrated that the peptide is up taken by viable embryos (mouse, cow, equine) [5]. In the case of bovine IVF embryos, FITC PIF uptake was determined at different stages, from blastocyst stage to the hatched stage, the ready to implant blastocyst. The data showed diffuse presence of FITC-PIF within the embryo. It also indicated that in the hatched region of the blastocyst, the uptake was most intense (i.e., polarized). Thus, in the extra villous trophoblast, PIF is at the leading edge of the impending direct embryo-maternal communication that takes place during implantation. The staining was specific since in the control, there was no stain; therefore further embryo viability was confirmed by using the DAPI stain. Further confirmation was provided since using a scrambled PIF (the PIF amino acid sequence in random order) was not up taken by the embryo. This strongly supported the view that PIF targets specific intracellular targets (i.e., receptor specific). The data in the equine embryo, another large mammal, helped to confirmed that the uptake of PIF by the embryo is also specific [9].

PIF: Binding Sites Lead to Biological Activity

There are specific binding sites in the developing embryo which upon binding lead to biological activity [9]. The next challenge was to identify the specific critical targets involved. For such an endeavor, several methods were employed. First, extracted mouse blastocysts were printed on a nitrocellulose-coated microarray and were probed with biotin-PIF identifying the different fractions by mass spectrometry. As a second validation method, specific PIF-based affinity column was used to isolate embryo tissue fraction and followed mass spectrometry which was used for identifying specific proteins. The leading proteins identified are protein-disulfide isomerase/thioredoxin (PDI/T) and heat shock proteins (HSPs), 14-3-3-scaffold proteins, tubulins, and actins. This indicated that PIF protects against oxidative stress and protein misfolding, which is critical for embryo survival and development. In addition, targeting structural proteins is important for visceral and neural development (also confirmed in later studies. Insulin-degrading enzyme (IDE) was identified earlier as a potential PIF binding partner with major role in neural and insulin regulation. Using anti-PIF antibody, the presence of this enzyme was demonstrated in the embryo extract as well. Similarly, the same method was used to identify a major PIF binding partner Kv1.3b-potassium channels relevant for potassium flux regulation and immune regulation.

Similar mechanisms may be operative both in the embryo and in diverse immune disorder models. Ancestral signals are already operative from the early embryo, and therefore their perturbation at an early stage of development can have significant consequences later in gestation. With respect to the embryo, it was important to establish that PDI/T, since it is expressed already in the earliest viable embryo reflecting its important role, can be targeted by PIF thus establishing functional relevance between PIF and the receptor protein. The data showed that in presence of a PDI/T inhibitor, the progress of cultured bovine embryos development to the blastocyst stage decreased fourfold. On the other hand, addition of PIF at the zygote stage and following embryo progress it increased more than twofold their reaching to the blastocyst stage [8]. This confirmed that PIF targets are specific receptors, and through them the peptide exerts its protective role. With respect to the mechanism involved, PIF can convert the receptor protein from oxidative to redox protective form. As such, PIF through its autonomous autotrophic and protective action enables the embryo to develop on its journey toward the uterus. This autonomy is very important when the maternal input is absent in vitro (IVF) and is also limited in vivo. The maternal role will become critical as implantation will take place.

PIF Primes and Promotes Uterine Receptivity

Viewing the embryo signaling through PIF action, it became important to examine whether PIF can prime the uterus prior to implantation. Such examination of PIF's effect on endometrial receptivity was carried out in four consecutive stages examining its effect on the human endometrium: starting prior to implantation, during implantation period (day 21), implantation decidua, and the first trimester decidua. The studies show diverging but synergetic effects, supporting maternal acceptance of the embryo as a continuum, since the requirements in each developmental stage are different. Progesterone secreted by the corpus luteum is the driver of the endometrium progress to the secretory-receptive phase [10–12].

Whether PIF alone can prime the endometrium as a sole agent was also tested. Prior to implantation, PIF increased beta-integrins expression in human epithelial cells. This is the first contact with the embryo while not affecting underlying stromal cells. The effect was similar when comparing the 9 and 15 AA version of PIF. At implantation window of human endometria, PIF promoted pro-receptive prolactin secretion. In human implantation decidua (HESC) (estrogen and progesterone induced), major effects on inflammation control, adhesive molecules, and antiapoptotic effect were observed, through genomic, proteomic, and pathway analysis [10].

The data revealed that over 500 genes were affected by exposure to low PIF concentration. The highest-ranking gene (53-fold increase) was IRAK1BP1—which is an IL receptor 1 associated kinase 1 BP1. It is involved in innate immune response to microbial agents. Other genes involved in the immune pathways included IL12RB which decreased ninefold. This is a receptor for a TH1 cytokine receptor and interferon-g secretion. The F506BP1 gene was also downregulated; this is a downstream immune response regulator. Increased CALD1 caldesmon1 which is an inhibitor of actin-myosin interaction, detrimental for the implantation process. In addition, both actin and myosin were identified as PIF targets, and STX3 syntaxin3 is involved in epithelial cells differentiation-decidua critical aspect for embryo attached. This is further enhanced by the increased DSCAML1, Down syndrome cell adhesion like 1 expression embryo adhesion pathway followed by increase observed in Sorbin and SORBS2 which interacts with the cytoskeleton and the increased connexin 45 expression which are critical for implantation involved in gap junction. On the other hand, BCL2 (B-cell CLL/lymphoma 2 downregulation of genes while increasing MDM2 expression limits proliferation is beneficial since decidualization, a differentiation process, favors embryo implantation.

Thus, this indicated PIF involvement in the apoptotic pathway required for an embryo embedding in the decidua. Whether PIF protects against excessive proliferation in HESC was determined. Following exposure to PIF, some growth factors increased while other decreased. Among them the increase in amphiregulin and epiregulin favor implantation, while those leading to proliferation as receptor ligand for EGF (betacellulin and IGF1) decreased. The data also showed that PIF action is exerted through the downregulation of the phosphorylated P-p38-MAPK/P-ERK1,2 and P-MEK-1 pathways. The enzyme catalytic site was not affected since addition of PIF to the purified enzyme did not block the enzyme activity [11].

Gene and protein expression are important; however, whether PIF also affected secreted products was also examined. In cultured HESC, PIF upregulated both mRNA and the secreted proinflammatory cytokine secretion GRO-a, MCP-3, and ICAM-1. This implies that the implantation milieu is inflammatory, which facilitates embryo ability to implant. This also demonstrates that the maternal receptive endometrium is not immune inert but an active process which takes place to enable implantation while maintaining an active immune surveillance through PIF action.

Finally, in the post-implantation phase, when embryogenesis is activated and is ongoing, PIF promoted trophic genes/proteins and those protecting against adverse environment. Similarly, the action of PIF on the FTDC, derived from first trimester decidua, a total of >500 genes were up- and downregulated [10]. As for the immune pathway, the T cell receptor (TRAalpha locus) was upregulated 60-fold—which transduces the effect of environmental agents on the decidua. Also, IRAKBP1 increased as well as TLR6 which interacts with TLR2 to mediate response to bacterial agents. This protein expression also was confirmed by western blot. Notably BCL2 decreased 27-fold, while the FAS ligand FAF-1 increased 21-fold, favoring the apoptotic pathway. The proteome analysis confirmed that macrophage inhibitory factor is increased. In addition, it showed PIF's regulatory effect on proteins peroxiredoxin and HSPB1, in line with binding and regulation of these proteins which protect against oxidative stress and protein misfolding as also shown in the embryo and human immune cells and in vivo. PIF also targets through specific receptors the decidua where regulation of the protein takes place. On the other hand, PIF decreased ERLIN 1, ER lipid raft associated 1, and NFLIX Nuclear factor 1 to non-detectable levels thereby protecting against neoplastic transformation and viral infection.

EGFs exert potent proliferative effects that would be detrimental for decidual formation. Indeed, PIF reduced the expression of this growth factor by 20-fold while increasing EPS15 which potentiates EGF degradation [11]. The pathway analysis showed similarity between effect of PIF on HESC and FTDC [11], by ranking (1) actin cytoskeleton related to remodeling of the decidua, (2) the integrin pathway which also shown in preimplantation embryo, and (3) G-protein-coupled receptors involved in embryo-maternal signaling. The genes related to xenobiotic metabolism were highly ranked in the FTDC, implying protection of the developing embryo against adverse environment. This is logical since at implantation, embryogenesis, which has not yet initiated, is the most vulnerable period.

As such, PIF has an integrated trophic/protective effect on the maternal system that starts with priming, without direct contact, and is continued during embryogenesis, the most critical period. The data also showed that maternal environment can be hostile and that adversity can affect embryos fate. The first structure that develops in the embryo is the notochord (nervous system), and by 5 weeks the neural tube is closed. Therefore, analyzing the effect of PIF on HESC and first trimester decidua revealed PIF's specific effect on several genes and proteins. In HESC the highest ranking was TLX2 which plays a major role in anterior brain differentiation. The increase in EPHA10 is involved in neural cells mobility. In contrast, RARA a retinoic acid receptor is known to promote teratogenicity of retinoic acid involved in growth arrest. In FTDC PIF promoted SMAD1,6 while decreasing Inhibin C, exerting both neurotrophic and neuroprotective effects. In HESC, PIF is involved in axonal signaling, neuroregulin, and synaptic long-term potentiation and preventing reduction of such signaling. In FTDC, axon guidance and signaling was dominant coupled with neurotropin and neuroregulin pathways [12].

As a confirmation of this therapeutic potential, the data is generated using PIF in the newborn and the adults where significant neuroprotection and neural repair are documented in relevant preclinical murine models (see below).

PIF Promotes Controlled Trophoblast Invasion: Intimate Embryo-Maternal Communication

The critical step in implantation is the ability of the trophoblast to invade and establish an intimate contact with the maternal system, a direct embryo-maternal communication. PIF is expressed in the human extra villous trophoblast (EVT), the first contact during invasion. Therefore, its role in trophoblast invasion was examined. The data showed that PIF promotes threefold trophoblast invasion in a dose-dependent manner [13]. This was determined using Matrigel invasion model and testing immortalized first trimester EVT (HTR-/SVneo cells). As mentioned before, EGF has proliferative effects on the decidua; therefore it was important to determine whether EGF could further potentiate the pro-invasive properties of PIF. Although EGF alone was effective, the combined administration did not further increase the EVT invasion. In contrast, the scrambled peptide had no effect, reflecting PIF's specific action. This promoted studies to examine whether similar observations can be made also with human first trimester EVT. When HLA-G expression (a prime tolerance molecule, described below) was examined in EVT, its expression was almost fivefold higher than in the villous trophoblast, indicating its tight interaction with the maternal environment [14].

PIF promotes human trophoblast invasion in a dose-dependent manner confirming the concentrations used with the transformed cells. On the other hand, PIF did not affect trophoblast proliferation. The invasion must be tightly regulated, adequate for the trophoblast to establish effective interaction with the endometrium but not to be excessive, which would be detrimental. The pathways and specific regulatory compounds involved require a delicate balance among pro-/anti-invasive markers. With respect to metalloproteinases, PIF increase MMP9 but not MMP2. MMP9 deficiency is shown to be involved in placental pathologies such as preeclampsia [13]. As for TIMP-1, PIF reduced its expression while again not affecting TIMP2. Finally, integrins both tested ITGAV and ITGA1 significantly increased but only at 24 and not 48 h preventing excessive invasion. The regulatory pathways involved were determined using specific inhibitors as shown also in the decidua. PIF action is dependent on the MAPK pathway, PI3Kinase, and Jak-Stat pathway. To determine mechanism of PIF action in detail, a global genomic analysis was carried out. By stringent analysis, a total of 146 genes were up- and downregulated [14]. Among them azurocidin was increased which has antibacterial activity and IL 17 F that has immune regulatory function as well as T cell receptor alpha variable 4. In contrast, PIF reduces RASL10A, RAS-like family10 A. Changes in several noncoding RNA genes were noted as well.

This pathway analysis revealed that PIF action relates to that of cancer, proliferation, cell death, and survival. Of great importance was to delineate that PIF action is dependent on p53, guardian of the genome. Specifically, effect on expression and the protein itself were determined in HTR-8/SVneo cells. The effect of PIF on trophoblast invasion was already documented in this transformed cell line [15]. Moreover, in EVT, data emerged that PIF increased pro-survival BCL2 and reduced Bax pro-apoptotic gene. To determine whether PIF antiapoptotic effect is dependent on p53 activity, HTR-8/SVneo cells were used showing that PIF reduced p53 phosphorylation. However, in presence of a specific inhibitor (TP53-specific siRNA), which decreased the protein by 90% blocked the PIF effect on the observed increase in BCL2 and the reduction in BAX expression. Finally, PIF also had an antiapoptotic effect since it blocked the effect of etoposide, an inhibitory topoisomerase II which induces p53-dependent apoptosis.

The determination of PIF's role in inducing tolerance for the embryo was studied using trophoblast cell line—JEG3, selected since it expresses HLA-G, a prime tolerance marker. The effect of PIF on this molecule was determined showing that it promotes HLA-G expression both surface and intracellular. In addition, PIF promoted HLA-C, E, and F complementing the pro-tolerance action [16]. The effect of progesterone was determined in a side-by-side comparison. The data showed that PIF's effect is superior as compared to the steroid determined by examining the effect on different HLAs. Moreover, PIF monotherapy was superior to progesterone administration also with respect to the effect on cytokines secretion. This steroid plays a major role in preparation of the endometrium for implantation. However, progesterone that is not derived from the corpus luteum and is generated in the placenta a late first trimester.

PIF is expressed starting from the two-cell stage embryo, and therefore its local action in the placenta is expected right away, whereas that of progesterone would occur much later. It was important to examine whether PIF supports progesterone's action in the trophoblast. The studies show that PIF promotes two aspects of progesterone. First, it increases the steroid secretion by the trophoblast. Therefore, it confirms that PIF facilitates progesterone expression and reflects a strong link between the two molecules. In other words, the first drives the other's production. Another aspect was testing PIF's effect on the trophoblast proteome, which showed that PIF upregulates the progesterone receptor; therefore it also potentiates the steroid's effect on the trophoblast. Finally, synergetic effect of PIF combined with progesterone was examined showing that the effect is complementary, opening the possibility that such a combination could be relevant in dealing with premature labor [16].

PIF Prevents Spontaneous and LPS-Induced Fetal Loss

The final step of these observations is whether improved embryo-maternal communication through PIF action can be translated to in vivo observations. To this end an immune intact model was used; thus eventual pregnancy loss would not be due to immune imbalance. Recognizing that spontaneous pregnancy loss in both human and murine is similar (~15%), it was important to determine whether PIF could reduce such a rate by improving embryo-maternal communication. Administration of PIF from conception showed that PIF reduced the rate threefold from 15% to 5% associated with normal progeny, evidenced by optimized weight gain of the fetus at birth [17]. Maternal inflammation has an adverse effect on the embryo; therefore PIF's effect was tested showing that following exposure to LPS bacterial antigen. A twofold reduction in fetal death and improved fetal, not placental, weight was evidenced.

Having shown that placental PIF expression declines at term, it was important to determine the expression of PIF also in murine gestation. Data from ex vivo experiments analyzing PIF expression from conception until late pregnancy showed that PIF is expressed in the placenta, and it is up taken by the uterine NK cells, of potential hostility to the fetus. This expression, however, decreases in late gestation as PIF is released from the UNK by granules in preparation for delivery. The murine study also showed that LPS upregulates PIF expression in the placenta; thus PIF, through its local regulatory action, may negate the adverse LPS action and thereby lead to increased fetal survival. Finally, PIF's protective action was examined showing that it is exerted on placental inflammasome NALP 3 where the reduction prevents apoptosis through reduced caspase 1 expression. This is coupled with a reduction in inflammatory cytokines such as IL18 that was shown elevated both in the placenta and circulation in patients with preeclampsia. Complementing the observation is the reduction in several circulating Th1 type cytokines which leads to the synergistic protection.

Considering that beyond fetal death in utero, prematurity is a major clinical problem, the effect of PIF was examined in a RPL model where PIF reduced LPS-induced premature delivery rate in immune intact mice fourfold. The question remained whether PIF action could also benefit the prematurely born fetus. Data showed that PIF protected the fetal brain by reducing microglial cells activation (iba1+ cells) and preserved neuronal cells migration (Cux-2 cells) while decreasing IL6 and INFg proinflammatory mediators. Thereby maternal prophylaxis with PIF prevents both fetal death and protect against prematurity while preventing fetal brain inflammation [18]. This further support data where PIF is shown effective reversing newborn brain damage induced by hypoxic ischemic encephalopathy [19, 20].

PIF Regulates Maternal Systemic Immunity

Beyond local immunity, maternal systemic immunity is an intricate part of reproductive success. It was important to analyze PIF and its communication with the maternal immune system [21]. The global (systemic) immune system needs to preserve homeostasis, and pregnancy clearly requires a major adaptation between embryo needs and maternal preservation. Mammalian reproduction is a very effective system. As mentioned above, genetic diversity and cross-species pregnancy are successful; therefore specific signaling that prevents embryo-maternal incompatibility must be operative and address global immunity.

For it to be effective, immune regulation must be global and start prior to implantation since it is well-known that the immune system through its pleiotropic effects will exploit any gap and will mount a powerful immune response. Whether PIF can exert such a role was shown in the endometrium through local inflammation since inflammation has to be present in a controlled manner for implantation to take place. Its effect on circulating cytokines also was shown above.

It became important to determine whether PIF also acts on the cellular immunity completing the embryo signal's integrated local and systemic effect. In line with this premise is that PIF targets the innate (macrophages/neutrophils) required to maintain basal immune defenses. The first is a first responder to pathogens and inflammation and acts through the antigen presenting system to regulate the adaptive part of immunity, T and B cells. The neutrophil through their direct antipathogen action addresses the pathogens aimed to neutralize them, bacteria. PIF binds the great majority of (CD14+) cells prior to pregnancy, while binding to CD3+ cells is low [22].

However, since pregnancy is associated with activated immune system (not suppression), PIF binding was shown to increase significantly, adapting to need [4]. The effect of PIF on global human and murine PBMC was examined. Exposure to mitogens, PHA, LPS, and PMA led to increased PIF binding to T (CD4, CD8) and B cells (CD19). The effect of PIF is exerted through binding principally to intracellular proteins with high homology that is found in the embryo [23]. The PDI/T and HSPs are mostly dominant in CD14+ cells, and binding to proteins involved in coagulation and immune regulatory and related proteins are more prominent in CD4+ and CD8+ cells. PIF action is minimal on basal immunity while reducing mixed lymphocyte reaction (related to tolerance) and activated cells where excessive proliferation decreased. This was coupled with a Th2/Th1 cytokine bias, while certain Th1 secretion is maintained to preserve antipathogen action reflecting perceived pathology. PIF action on hyperactivated cells was determined; antiCD3/CD28 antibody induced PBMCs activation. Such an exposure led to major large-scale changes in immune response where effect after 24 h exposure was different at 48 h; both increase and decrease gene expression.

Thus, PIF action is dynamic and can respond to challenge. In examining the dynamics of PIF's effect on the cytokine profile, a similar pattern emerged. PIF effect started after 6 h, plateaued at 48 h, and returned to the baseline by 96 h [22]. As recently demonstrated, PIF binds and regulates potassium channels Kv1.3b which turns out to be the binding site of cortisone. PIF reduces K+ flux while not affecting Ca++ flux in the cells; thus immune modulation takes place, while suppression is avoided [4]. It is recognized that regulatory T cells increase following conception, which is considered important for maternal tolerance. PIF is shown to target those CD4+/CD25+/FoxP3+ cells; thus it is involved in tolerance induction on the systemic level as well [23]. It was mentioned that PIF is up taken by uterine NK cells, a potential protection against maternal hostility. As part of effective embryo-maternal communication, systemic NK cell cytotoxicity must be mitigated; otherwise, if these cells are overactive and are increased in the maternal circulation, pregnancy may fail repeatedly. PIF's effect on NK cells cytotoxicity was determined in a large cohort of patients with recurrent pregnancy loss [24]. The data showed that PIF reduced such cytotoxicity, reflecting a protective effect. The effect is indirect through global immune effects since the binding to NK is low and is not increased following activation by PHA, a potent mitogen [22]. The protective mechanism involved in PIF action is through reduction of CD69 expression, a major inducer of inflammation [23]. Beyond the in vitro studies mentioned above, PIF's regulatory effect on the immune system targeting macrophages and neutrophils in vivo was also demonstrated outside pregnancy complementing that information.

Concluding Remarks

Pregnancy is a highly complex interplay that nature successfully designed. Herein the aim is to dissect the contribution of the embryo through effective embryo-maternal communication. It can be asserted that the embryo is the driver and the maternal system is the responsive entity. Maternal awareness of the presence of the embryo starts very early, and the signaling associated with its success must be cross-mammalian, since early stages of development are highly similar from mouse to human.

Our search identified this cross-species, evolutionary-conserved, function-maintaining pivot signaling and called it the preimplantation factor. Insofar, detection in human but also mouse, bovine, porcine, equine, and sheep and in primates was evidenced. PIF protein receptors were similarly shown in mouse embryos, human immune cells, mouse brain, human liver, and kidney. Thus, the binding sites and associated pathways involved are also well preserved across mammalian species.

The embryo exerts three intercalating actions: protection against oxidative stress, immune regulation, and regeneration. PIF has been shown to be intimately involved, leading to an effective communication with the mother in a certain degree guiding her on the path to benefit both parties. The embryo is a semi-allogeneic product (partly derived from the maternal genome) or, alternatively, a willing partner with no long-term benefit in donor and cross-species gestation. Although she is not carrying her genetic makeup, post-delivery, she benefits from the newborn and continues to nurture throughout life.

Those lessons learnt on PIF's role in the embryo original stem cells and effective communicating it established with the maternal organism are being translated in outside pregnancy setting.

Synthetic PIF shown effective in diverse preclinical immune disorders and transplant [25–35], and the safety established in FDA-directed toxicology studies enables completion of the first-in-human FDA Fast-Track Phase I clinical trial, in patients with autoimmune disease, and enables progress to Phase II trial in both pregnancy and non-pregnant patients.

References

1. Barnea ER, et al. The first twelve weeks of gestation. New York: Springer-Verlag; 1992. p. 544.
2. Barnea ER. Applying embryo-derived immune tolerance to the treatment of immune disorders. Ann N Y Acad Sci. 2007;1110:602–18.
3. Barnea ER, et al. Progress in characterization of pre-implantation factor in embryo cultures and in vivo. Am J Reprod Immunol. 1999;42(2):95–9.
4. Barnea ER, et al. PIF direct immune regulation: blocks mitogen-activated PBMCs proliferation, promotes TH2/TH1 bias, independent of Ca(2+). Immunobiology. 2015;220(7):865–75.
5. Calix R, Ornaghi S, Wilson J, Fernandez N, Vialard F, Barnea ER, Paidas MJ. PIF and endocrinology of implantation and establishment of early pregnancy: a contemporary view. Pediatr Endocrinol Rev. 2017;15(2):147–58.
6. Stamatkin CW, et al. PreImplantation Factor (PIF) correlates with early mammalian embryo development-bovine and murine models. Reprod Biol Endocrinol. 2011;9:63.
7. Stamatkin CW, et al. Preimplantation factor negates embryo toxicity and promotes embryo development in culture. Reprod Biomed Online. 2011;23(4):517–24.
8. Goodale LF, et al. PreImplantation factor (PIF) protects cultured embryos against oxidative stress: relevance for recurrent pregnancy loss (RPL) therapy. Oncotarget. 2017;8(20):32419–32.

9. Barnea ER, et al. Insight into PreImplantation Factor (PIF*) mechanism for embryo protection and development: target oxidative stress and protein misfolding (PDI and HSP) through essential RIKP binding site. PLoS One. 2014;9(7):e100263.
10. Paidas MJ, et al. A genomic and proteomic investigation of the impact of preimplantation factor on human decidual cells. Am J Obstet Gynecol. 2010;202(5):459 e1–8.
11. Barnea ER, Kirk D, Paidas MJ. Preimplantation factor (PIF) promoting role in embryo implantation: increases endometrial integrin-alpha2beta3, amphiregulin and epiregulin while reducing betacellulin expression via MAPK in decidua. Reprod Biol Endocrinol. 2012;10:50.
12. Duzyj CM, et al. PreImplantation Factor (PIF*) promotes embryotrophic and neuroprotective decidual genes: effect negated by epidermal growth factor. J Neurodev Disord. 2014;6(1):36.
13. Duzyj CM, et al. Preimplantation factor promotes first trimester trophoblast invasion. Am J Obstet Gynecol. 2010;203(4):402 e1–4.
14. Moindjie H, et al. Preimplantation factor (PIF) promotes human trophoblast invasion. Biol Reprod. 2014;91(5):118.
15. Moindjie H, et al. PreImplantation Factor is an anti-apoptotic effector in human trophoblasts involving p53 signaling pathway. Cell Death Dis. 2016;7(12):e2504.
16. Hakam MS, et al. Preimplantation factor (PIF) promotes HLA-G, –E, -F, -C expression in JEG-3 choriocarcinoma cells and endogenous progesterone activity. Cell Physiol Biochem. 2017;43(6):2277–96.
17. Di Simone N, et al. Synthetic PreImplantation Factor (PIF) prevents fetal loss by modulating LPS induced inflammatory response. PLoS One. 2017;12(7):e0180642.
18. Spinelli M, et al. Synthetic preimplantation Factor (spIF) reduces inflammation and prevents preterm birth. PLoS One. 2020;15(6):e0232493. https://doi.org/10.1371/journal.pone.0232493.
19. Mueller M, et al. PreImplantation factor promotes neuroprotection by targeting microRNA let-7. Proc Natl Acad Sci U S A. 2014;111(38):13882–7.
20. Mueller M, et al. PreImplantation Factor bolsters neuroprotection via modulating Protein Kinase A and Protein Kinase C signaling. Cell Death Differ. 2015;22(12):2078–86.
21. Barnea ER, et al. Immune regulatory and neuroprotective properties of preimplantation factor: from newborn to adult. Pharmacol Ther. 2015;156:10–25.
22. Barnea ER, et al. PreImplantation factor (PIF*) regulates systemic immunity and targets protective regulatory and cytoskeleton proteins. Immunobiology. 2016;221(7):778–93.
23. Sbracia M, et al. PreImplantation Factor in endometriosis: a potential role in inducing immune privilege for ectopic endometrium. PLoS One. 2017;12(9):e0184399.
24. Roussev RG, et al. Preimplantation factor inhibits circulating natural killer cell cytotoxicity and reduces CD69 expression: implications for recurrent pregnancy loss therapy. Reprod Biomed Online. 2013;26(1):79–87.
25. O'Brien C, et al. Randomized, double-blind, placebo-controlled, single ascending dose trial of synthetic preimplantation factor in autoimmune hepatitis. Hepatol Commun. 2018;2:1232–43.
26. Migliara G, et al. PIF* promotes brain re-myelination locally while regulating systemic inflammation-clinically relevant multiple sclerosis M. smegmatis model. Oncotarget. 2017;8(13):21834–51.
27. Weiss L, et al. Preimplantation factor (PIF) analog prevents type I diabetes mellitus (TIDM) development by preserving pancreatic function in NOD mice. Endocrine. 2011;40(1):41–54.
28. Weiss L, et al. Preimplantation factor (PIF*) reverses neuroinflammation while promoting neural repair in EAE model. J Neurol Sci. 2012;312(1–2):146–57.
29. Chen YC, et al. PreImplantation factor prevents atherosclerosis via its immunomodulatory effects without affecting serum lipids. Thromb Haemost. 2016;115(5):1010–24.
30. Azar Y, et al. Preimplantation factor reduces graft-versus-host disease by regulating immune response and lowering oxidative stress (murine model). Biol Blood Marrow Transplant. 2013;19(4):519–28.
31. Shainer R, et al. PreImplantation factor (PIF) therapy provides comprehensive protection against radiation induced pathologies. Oncotarget. 2016;7(37):58975–94.

32. Balyura M, et al. PreImplantation Factor (PIF*) regulates stress-induced adrenal steroidogenesis and anti-inflammatory cytokines: potential application for bioartificial adrenal transplant. Horm Metab Res. 2017;50:168–74.
33. Hayrabedran S, et al. The core sequence of PIF competes for insulin/amyloid β in insulin degrading enzyme—potential treatment for Alzheimer's disease. Oncotarget. 2018;
34. Feichtinger M, et al. Allogeneic ovarian transplantation using immunomodulator preimplantation factor (PIF) as monotherapy restored ovarian function in olive baboon. J Assist Reprod Genet. 2017;35(1):81–9. https://doi.org/10.1007/s10815-017-1051-y.
35. Margoulis D, et al. sPIF promotes myoblast differentiation and utrophin expressin while inhibiting fibrosis in Duchenne muscular dystrophy via the H19/miR-675/let-7 and miR21 pathways. Cell Death Dis. 2019;10(2):82.

Part V

Some Difficult Patients

Management of Infertility in Overweight or Obese Polycystic Ovary Syndrome Patients

19

Xiangyan Ruan, Yu Yang, Muqing Gu, and Pooja Dhungel

Prevalence

Polycystic ovary syndrome (PCOS) is the most common reproductive endocrine disease in women of reproductive age and the most common cause of anovulatory infertility [1], accounting for about 70%~80% of anovulatory infertility. According to the report from Department of Gynecological Endocrinology, Beijing Obstetrics and Gynecology Hospital, Capital Medical University, PCOS is one of the most important diseases. More than 500 outpatients visit the department daily, among them at least 50% are diagnosed with PCOS, and more than 50,000 per year of our PCOS patients get treatment as described in this chapter. The disease can begin in early adolescence, the etiology is not yet clear, the pathogenesis is complex, and it is related to environmental factors (especially nutrition). Particularly, genetical factors may also play an important role in the development of the disease.

Obesity is an increasingly serious global health problem, about 25% of women are overweight before pregnancy [1]. But PCOS patients are more likely to be obese. The prevalence of obesity in PCOS is about 2.8 times higher than that in non-PCOS [2], and 70~80% of PCOS patients have overweight and abdominal obesity [1]. Proportion of PCOS with obesity varies among different races; it is significantly higher in white women than Asian women [3]. Both obesity and PCOS have adverse effects on metabolism. Compared with non-obese PCOS or non-PCOS obesity, obese PCOS has more serious glucose and lipid metabolism disorders [4]. In PCOS patients, obesity, especially visceral fat, increases insulin resistance (IR), resulting

X. Ruan (✉) · Y. Yang · M. Gu · P. Dhungel
Department of Gynecological Endocrinology, Beijing Obstetrics and Gynecology Hospital, Capital Medical University, Beijing Maternal and Child Health Care Hospital, Beijing, People's Republic of China
e-mail: ruanxiangyan@ccmu.edu.cn

J. G. Schenker et al. (eds.), *Hot Topics in Human Reproduction*, Reproductive Medicine for Clinicians 3, https://doi.org/10.1007/978-3-031-24903-7_19

in compensatory hyperinsulinemia, resulting in as follows [5]: (1) reduced fat decomposition and increased synthesis; (2) increased sensitivity of follicular membrane cells to LH, resulting in increased androgen synthesis; and (3) increased inflammatory adipokines. These factors will further aggravate IR, increase visceral fat, and cause vicious circle. This chapter will focus on comprehensive management of obese/overweight PCOS, to break this vicious cycle, further improve metabolism, recover ovulation, achieve pregnancy and live birth, and reduce adverse pregnancy outcomes and pregnancy complications.

Pay Attention to Visceral Fat, Not Only BMI

At present, the body mass index (BMI) is most commonly used to measure whether obese or overweight. The World Health Organization (WHO) defines BMI between 25.0 kg/m^2 and 29.9 kg/m^2 as overweight and ≥30 kg/m^2 as obesity. In China, BMI ≥24.0 kg/m^2 is regarded as overweight, and BMI ≥ 28.0 kg/m^2 is defined as obesity in adults. Although it is easy to get BMI, BMI calculation only uses body weight and does not further distinguish the weight of muscle and adipose. Therefore, even if the adipose tissue increased, a part of the obese population can be omitted due to decreased muscle. Whether obese PCOS or non-obese PCOS, compared with age and BMI-matched non-PCOS, visceral fat and visceral fat index are increased [6]. Even PCOS women with normal BMI have higher visceral fat and inflammatory markers than non-obese controls [7]. Therefore, the determination of obesity cannot only be based on body weight and BMI, but also should pay attention to fat content, especially visceral fat content. Visceral fat is related to the increased risk of cardiovascular disease. If only BMI is concerned and visceral fat is ignored, the management of population with normal BMI but excessive fat will be omitted, and we cannot do early prevention of long-term cardiovascular complications in this population.

Why Is PCOS Infertile?

1. Ovulation disorders

   PCOS is the most common cause of anovulatory infertility. In PCOS, follicular stagnates at antral follicle stage and cannot develop into mature follicles and ovulation, resulting in oligo-ovulation or anovulation.
2. Poor quality of oocytes and embryos

   In recent years, high-throughput sequencing has developed rapidly. Genome data shows that the transcriptome characteristics of oocytes in PCOS are different from non-PCOS, which involve meiosis, oxidative stress in follicles, and glucose and lipid metabolism regulation [8]. The increase in oxidative stress will lead to excessive production of reactive oxygen species, thereby increasing the incidence of meiotic abnormalities and ultimately reducing oocyte quality.

Obesity is an important cause of low oocyte quality. Even with regular ovulation, increased BMI is associated with lower pregnancy rates, increased risk of pregnancy loss, and early abortion [1]. In morphology, the frequency of centrally located granular cytoplasm of oocytes in overweight and obese women was higher than that in normal weight women [9]. Preimplantation genetic diagnosis showed that the centrally located granular cytoplasm of oocytes could lead to the increase of aneuploid embryos [10]. In follicular microenvironment, insulin, lactic acid, triglycerides (TG), and C-reactive protein (CRP) levels were elevated in women with higher BMI. The increase of TG in follicular fluid has lipotoxic effect on oocytes, resulting in increased lipid content in oocytes, increased endoplasmic reticulum stress, and abnormal nuclear maturation [8]. Increased CRP suggests that oxidative stress is a possible mechanism of obesity affecting oocyte quality.

3. Damaged endometrial receptivity

   Good endometrial receptivity is one of the necessary conditions for oosperm localization, adhesion and invasion, and subsequent blastocyst division and embryonic development [11]. Compared with healthy control, PCOS have a series of adverse pregnancy outcomes, such as low embryo implantation rate and high abortion rate. Impaired endometrial receptivity may be an important reason for these adverse pregnancy outcomes in PCOS. Studies have shown that estrogen and progesterone secretion disorders, high androgen, hyperinsulinemia, obesity, and low chronic inflammatory are involved in impaired endometrial receptivity in PCOS [11]. Abnormal expression of estrogen, progesterone, and androgen receptors and their modulators in endometrium are found in PCOS. Compared with normal women of childbearing age, endometrial receptivity in PCOS patients may be impaired by the following mechanisms [11]: (1) Because the endometrium is stimulated by estrogen for a long time without antagonistic effect of progesterone, the expression of estrogen receptor activator AIBI and transcription factor TIF2 in the endometrium of PCOS is significantly increased in secretory phase, which further activates estrogen receptor α, enhances the long-term proliferative effect of estrogen on endometrium, causes progesterone resistance, and affects the establishment of endometrial receptivity. (2) A large number of studies have confirmed that there are not only increases in inflammatory factors such as interleukin 6 (IL-6) and CC motif ligand (CCL2) in proliferative endometrium; uterine natural killer (Unk) cells in post-secretory endometrium is also significantly reduced [12]. The above can destroy the normal immune system of endometrium, which may be one of the key factors for damaged endometrial receptivity in PCOS patients.

Pregnancy Outcome and Offspring Health

Compared with non-PCOS women, PCOS had two to four times higher risk of miscarriage, 2.8~3.7 times higher risk of gestational diabetes, and three to four times higher risk of hypertensive disorder complicating pregnancy (including gestational hypertension and preeclampsia) [13]. Studies also found that the risk of premature delivery in PCOS is higher than that in non-PCOS, but it is not clear whether the increase of this risk is due to early spontaneous delivery or due to maternal or fetal diseases that require artificial early termination of pregnancy [13, 14].

The pregnancy complications of PCOS may interact with some risk factors of PCOS, further deteriorating the outcomes of fetuses and offspring [13]. Various adverse outcomes of fetuses and newborns can be seen in PCOS [13], such as large for gestational age, small for gestational age infant, fetal intrauterine growth restriction, low birth weight infant, fetal macrosomia, etc. In addition, the risk of PCOS offspring entering neonatal intensive care unit and perinatal mortality also increased [14]. The anogenital distance [15] and sebum [16] secretion in newborn girls of PCOS women increased, which indicates that the fetus of PCOS is exposed to androgen in the uterus. The influence of PCOS on offspring not only may be in the intrauterine and neonatal period but also may have adverse effects on the long-term health of offspring. The risk of PCOS girls being diagnosed as PCOS was five times higher than that of non-PCOS women [17]. Animal study showed that prenatal androgen exposure led to reproductive and metabolic dysfunction in female offspring, and the effects of this androgen exposure even had transgenerational effects, lasting to F3 generations [17]. The latest studies suggested that serum AMH level in PCOS during pregnancy was higher than that in non-PCOS [18]. Animal study in mice have shown that high AMH exposure during pregnancy can also lead to PCOS-like reproductive and neuroendocrine phenotypes in adult female offspring [19].

Management of Overweight/Obese PCOS

Lifestyle Improvement

Lifestyle improvement is the first-line treatment for overweight/obese PCOS. Improving lifestyles before pregnancy is particularly important, which can reduce body weight, improve IR, reduce androgen levels, improve ovulation and hirsutism, and increase ovarian responsiveness to gonadotropins during ovulation induction and assisted reproduction. Generally, weight loss 5~10% can improve PCOS [5]. For women without ovulation, weight loss >5% can restore ovulation.

Diet

Mediterranean diet (MedDiet) is recognized as the most healthy diet because of its unique characteristics, including regular consumption of unsaturated fat, fiber, low-glucose index (GI) carbohydrates, antioxidants, vitamins, and an appropriate amount of animal-derived protein, which can reduce inflammation and oxidative stress markers and improve lipid profile and insulin sensitivity, thereby reducing the risk of chronic diseases such as obesity, IR, and type II diabetes [20]. MedDiet is also considered a primary prevention of metabolic syndrome. Considering the close relationship between PCOS and obesity, chronic low inflammation, IR, and the benefits of MedDiet in metabolic syndrome and anti-inflammatory, MedDiet may be a good choice for PCOS. Nurse Health Study II [1] is a prospective cohort study based on a food frequency questionnaire. The reproductive outcomes of 17,544 nurses aged 25~42 years were analyzed. The results showed that the risk of infertility caused by ovulation disorders and other causes was reduced by 66% and 27%, respectively, for nurses following the MedDiet. Another study showed that the closer the dietary pattern was to the MedDiet, the lower CRP, HoMA-IR, testosterone, and Ferriman-Gallwey scores in PCOS patients [21].

Ketogenic diet (KD) is a dietary pattern that induces nutritional ketosis through high-fat intake and strict restriction of carbohydrate intake [20]. About 70% of energy is supplied by fat and about 5% by carbohydrate. Due to the lack of carbohydrate supply in the body, instead of burning fat to produce energy to achieve weight loss, KD was originally used to treat refractory epilepsy. In recent years, KD has become a hot topic in obesity and PCOS. KD can reduce postprandial insulin, weight, and body fat and improve IR [20]. In addition, in overweight/obese PCOS, KD intervention can also decrease LH/FSH, free testosterone, and dehydroepiandrosterone sulfate, increase sex hormone binding globulin, and improve menstrual cycle [22]. Although current studies have shown that KD has a promising effect on metabolism and hormone improvement, there are still some problems to be considered: (1) It's hard to insist for a long time. Because KD needs to strictly limit the intake of carbohydrates, which will make people feel unhappy. Halitosis, constipation and diarrhea, muscle spasm, headache, vitamin deficiency, kidney stones, and other side effects also make KD easy to interrupt [23]. Up to now, the intervention time of KD research is almost no more than 3 months. Although a study intervened for 6 months, 11 patients were included, and only five patients completed the intervention for 6 months [24]. So KD management is hard to implement for a long time. (2) Long-term safety: the energy source of KD is mainly fat, and the safety of long-term high-fat intake and elevated ketone body is a problem that must be considered. Because the present research is almost no more than 3 months, it is impossible to determine the long-term safety of KD. PCOS is a chronic disease that needs long-term management [25]. KD may help PCOS patients reduce weight and improve IR in the short term. However, KD is not an ideal solution for long-term management of PCOS due to above problems.

Although many studies have shown that various dietary patterns had good effects on obese PCOS in reducing weight and visceral fat, improving IR, reducing androgen, and restoring ovulation, there is no unified lifestyle suitable for all patients. Due to the high heterogeneity of PCOS, lifestyle intervention should be individualized. In the clinic of Beijing Obstetrics and Gynecology Hospital, Capital Medical University, after a comprehensive analysis of diet, exercise, living habits, environmental factors, body composition, motor function, and static metabolic rate of each PCOS patient, the intake of carbohydrate, fat, protein, and trace elements is adjusted accordingly, rather than only limiting the intake of energy and carbohydrate. Through this individualized nutritional guidance, almost all of our patients can adhere to a reasonable diet for a long time, so as to improve metabolism and fertility.

Exercise

The most favorable exercise for PCOS is still inconclusive. The international evidence-based guidelines for PCOS management suggest that for overweight or obese PCOS patients, in order to achieve moderate weight loss, prevent weight rebound, and greater health benefits, it is recommended to have at least 250 min of moderate intensity activity or 150 min of vigorous intensity activity per week or an equivalent combination of both, as well as 2 days of discontinuous muscle strengthening activities involving major muscle groups per week [26]. For the choice of exercise mode, overweight/obese patients should try to choose exercise with small load on knee joint, such as swimming and biking on the flat road.

Dietary Supplement

Most PCOS patients have dietary imbalance, such as insufficient intake of fiber, omega-3 fatty acids, calcium, magnesium, zinc, and vitamins (folic acid, vitamin C, B12, and vitamin D) [27]. Appropriate nutritional supplements have a positive effect on improving PCOS. Inositol belongs to vitamin B (vitamin B8), and there are nine isomers. The most common are muscle inositol (MI) and D-chiral inositol (DCI), which are in dynamic balance in healthy human. In the ovaries of PCOS patients, MI and DCI are imbalanced, with decreased MI and increased DCI [28]. DCI is a kind of aromatase inhibitor. Excessive DCI concentration limits the transformation of androgen to estrogen, resulting in high androgen, affecting the development and maturation of follicular and oocyte quality. Inositol is a safe and effective nutritional supplement which can improve PCOS. Up to now, most studies support that MI combined with DCI at a dose of 40:1 is the most effective treatment regimen, which can improve IR and hyperandrogenism, restore spontaneous ovulation, and improve the fertility potential in PCOS [29, 30]. In addition, appropriate supplementation of vitamin D, zinc, and omega-3 fatty acids can also help improve PCOS.

In recent years, animal and clinical studies have shown that changes in the intestinal flora are closely related to PCOS and metabolic syndrome [31–33]. Regulation of intestinal flora may be a potential direction for the treatment of PCOS. WHO defines probiotics as live microorganisms beneficial to the host when consumed in sufficient quantities [34–37]. Prebiotics refer to those that are not absorbed by the host but can selectively promote the growth of beneficial bacteria in the body [34–37]. Synbiotics are dietary supplements composed of probiotics and prebiotics [34–37]. Studies have shown that 12 weeks after treatment with probiotics/prebiotics/synbiotics in PCOS patients, weight and waist circumference can be reduced, and metabolism can be improved, including reduced IR, triglycerides, low-density lipoprotein, and cholesterol [36, 38]. Improvement of chronic inflammatory state can also be achieved, such as decreased CRP level [39]; probiotics/prebiotics/synbiotics can also improve reproduction, including increased sex hormone-binding globulin and reduced androgen level and restoring menstrual cycle [38, 40]. From the existing research results, probiotics/probiotics/synbiotics have a certain role in the treatment of PCOS.

Pharmacologic Therapy to Lose Weight

It is difficult to lose weight to ideal weight by improving lifestyle through adjusting diet and exercise, and pharmacologic therapy can be used to help weight loss. Metformin, liraglutide, and orlistat are the most commonly used drugs for weight loss in PCOS. Metformin improves insulin sensitivity by increasing peripheral glucose uptake and utilization, reducing glucose production in liver and reducing intestinal glucose absorption [5]. Liraglutide is a long-acting glucagon-like peptide-1(GLP-1) receptor agonist that binds to endogenous GLP-1 receptors, stimulates insulin secretion, and inhibits central appetite [5]. Orlistat plays a pharmacological role in the gastrointestinal tract, inhibiting the lipase, further preventing the hydrolysis of triglycerides into free fatty acids, and reducing the intestinal absorption of triglycerides in the diet [5]. A network meta-analysis [41] showed that the effects of liraglutide, orlistat, and metformin on weight loss decreased in turn. Metformin combined with liraglutide has the best weight loss effect.

Ruan's team [42, 43] compared the body fat improvement after 12 weeks of treatment with ethinylestradiol/cyproterone acetate (EE/CPA) alone, EE/CPA + metformin, EE/CPA + orlistat, and EE/CPA + metformin + orlistat. The results showed that EE/CPA + orlistat had the best effect on reducing weight and body fat, decreasing androgen, and improving glucose metabolism and the least adverse effect. Orlistat could reduce triglyceride levels after treatment for 3 months, which was better than EE/CPA alone and EE/CPA + metformin group. Because orlistat reduces the absorption of fat, the absorption of fat-soluble vitamins can also be reduced, so pay attention to the supplement of fat-soluble vitamins during orlistat treatment.

In addition to improving IR in PCOS patients, metformin can also reduce androgen levels [44] and increased menstrual frequency [44] and reduce the incidence of ovarian hyperstimulation syndrome [45] in assisted reproductive technology. Metformin use during pregnancy in PCOS may also be beneficial. A randomized, double-blind, multicenter placebo-controlled trial (PregMet study) [46] showed that taking metformin from early pregnancy to delivery can reduce the risk of early abortion, late abortion, and premature delivery. A latest population-based cohort study [47] based in Swedish population showed that PCOS in women without metformin use during pregnancy was associated with higher risks of preeclampsia (OR = 1.09, 1.02–1.17), gestational diabetes (OR = 1.71, 1.53–1.91) and caesarean section (OR = 1.08, 1.04–1.12), preterm birth (OR = 1.30, 1.23–1.38), low birth weight (OR = 1.29, 1.20–1.38), low Apgar scores (OR = 1.17, 1.05–1.31), and large for gestational age (OR = 1.11, 1.03–1.20). However, metformin can penetrate the placental barrier, and recent studies have found that intrauterine metformin exposure may have a long-term impact on the health of offspring. The follow-up results of

PregMet study [48] showed the weight and BMI Z score of the offspring at the age of 4 in the metformin group were higher than those in the placebo group. The proportion of overweight/obesity of offspring in metformin group was also higher than that in placebo group (32% vs 18%). The follow-up time was further extended to 5~10 years old. It was found that the BMI Z score of PCOS offspring in intrauterine metformin exposure group was still higher than placebo group and more children with central obesity [49]. The increased BMI of PCOS offspring exposed to metformin during pregnancy may indicate an increased risk of heart and metabolic diseases in this population in the future. Animal study [50] found that intrauterine metformin exposure even caused declined fertility in adult male mice. Adult male mice exposed to metformin during fetal period had increased proportion of abnormal sperm head, DNA damage, and decreased sperm quality, resulting in a 30% reduction in the litter size compared with the control group. Metformin is a commonly used drug for the treatment of gestational diabetes, which can reduce the risk of some adverse pregnancy outcomes and pregnancy complications of PCOS. However, in recent years, more and more studies suggested that metformin exposure during pregnancy may affect the long-term health of offspring, we should be more cautious when treating PCOS pregnant with metformin.

Bariatric Surgery

In morbid obesity PCOS women who fail to lose weight through diet, exercise, and pharmacologic intervention, bariatric surgery is one of the last sequential recommendations for weight loss intervention and metabolic improvement in PCOS, but has considerably greater complexity [5]. Bariatric surgery is the last choice for patients with extremely obese PCOS (BMI > 40 kg/m^2) who have failed lifestyle and medication intervention [5]. It is not enough to lose weight by 5~10% for such extremely obese PCOS, and it needs to lose weight by 25~50% to improve PCOS [5]. A meta-analysis [51] of the efficacy of weight loss surgery in patients with extremely obese PCOS showed that after weight loss surgery, the incidence of abnormal menstruation and hairiness was significantly reduced. Bariatric surgery can also reduce serum total and free testosterone levels and decrease the risk of type II diabetes and hypertension. When considering the benefits of bariatric surgery for morbid obese PCOS patients, the related risks must also be weighed. In addition to short-term postoperative complications [52] such as bleeding, wound infection, urinary tract infection, and venous thromboembolism, some long-term risks need to be considered, for example, increased risk of new depression [53], vitamin deficiency [54], secondary hyperparathyroidism [55], and fracture [54].

Losing weight too quickly and too much or excessive exercise will cause functional hypothalamic amenorrhea (FHA). To prevent FHA in the process of weight loss in overweight/obese PCOS, limitation of energy intake or exercise should not be excessive. Cholesterol is the raw material for human sex hormone synthesis. Low body fat leads to insufficient raw materials for sex hormone synthesis, unable to produce enough estradiol which results in amenorrhea. To maintain regular menstrual cycles, body fat needs to reach 22% [56]. When energy storage cannot meet the energy needs of the body, the downregulation of hypothalamic-pituitary-ovary (HPO) axis and the decrease of gonadotropin-releasing hormone (GnRH) secretion lead to the decrease of follicle stimulating hormone (FSH), luteinizing hormone (LH), follicular development, and estrogen secretion and also lead to amenorrhea [56].

Combined Oral Contraceptives (COC)

For women without fertility desire, COC help establish regular artificial cycles, reduce androgen, improve metabolism, and prevent long-term complications such as endometrial cancer. For women with fertility needs, studies have shown that PCOS patients taking ethinylestradiol and cyproterone acetate three cycles before ovulation induction therapy can reduce the risk of gestational diabetes, gestational hypertension, and premature delivery in PCOS patients [57].

Summary

PCOS patients are more likely to be obese, about 70~80% of PCOS with overweight or abdominal obesity. Obesity, especially visceral fat, aggravates metabolic and endocrine disorders in PCOS. Lifestyle improvement is the first-line treatment for overweight/obese PCOS. Comprehensive management of diet, exercise, dietary supplements, and COC before pregnancy can reduce weight, improve IR, reduce androgen levels, restore reproductive function, increase sensitivity to gonadotropin, and reduce adverse pregnancy outcomes. All PCOS patients should be recommended for lifestyle improvement.

Acknowledgments Supported by Beijing Municipal Administration of Hospitals' Ascent Plan (DFL20181401).

References

1. Garruti G, Depalo R, De Angelis M. Weighing the impact of diet and lifestyle on female reproductive function. Curr Med Chem. 2019;26(19):3584–92.
2. Chudzicka-Strugała I, Kubiak A, Banaszewska B, et al. Effects of synbiotic supplementation and lifestyle modifications on women with polycystic ovary syndrome. J Clin Endocrinol Metabol. 2021;106(9):2566–73.
3. Lim SS, Davies MJ, Norman RJ, et al. Overweight, obesity and central obesity in women with polycystic ovary syndrome: a systematic review and meta-analysis. Hum Reprod Update. 2012;18(6):618–37.
4. Shi W, Zhao Q, Zhao X, et al. Analysis of endocrine and metabolic indexes in non-obese patients with polycystic ovary syndrome and its compare with obese patients. Diabetes Metab Syndr Obes. 2021;14:4275–81.

5. Glueck CJ, Goldenberg N. Characteristics of obesity in polycystic ovary syndrome: etiology, treatment, and genetics. Metabolism. 2019;92:108–20.
6. Jena D, Choudhury AK, Mangaraj S, et al. Study of visceral and subcutaneous abdominal fat thickness and its correlation with cardiometabolic risk factors and hormonal parameters in polycystic ovary syndrome. Indian J Endocrinol Metab. 2018;22(3):321–7.
7. Durmus U, Duran C, Ecirli S. Visceral adiposity index levels in overweight and/or obese, and non-obese patients with polycystic ovary syndrome and its relationship with metabolic and inflammatory parameters. J Endocrinol Investig. 2017;40(5):487–97.
8. Palomba S. Is fertility reduced in ovulatory women with polycystic ovary syndrome? An opinion paper. Hum Reprod. 2021;36(9):2421–8.
9. Depalo R, Garruti G, Totaro I, et al. Oocyte morphological abnormalities in overweight women undergoing in vitro fertilization cycles. Gynecol Endocrinol. 2011;27(11):880–4.
10. Kahraman S, Yakin K, Dönmez E, et al. Relationship between granular cytoplasm of oocytes and pregnancy outcome following intracytoplasmic sperm injection. Hum Reprod. 2000;15(11):2390–3.
11. Bai X, Zheng L, Li D, et al. Research progress of endometrial receptivity in patients with polycystic ovary syndrome: a systematic review. Reprod Biol Endocrinol. 2021;19(1).
12. Piltonen TT. Polycystic ovary syndrome: endometrial markers. Best Pract Res Clin Obstet Gynaecol. 2016;37:66–79.
13. Bahri KM, Boyle JA, Tay CT, et al. Polycystic ovary syndrome and adverse pregnancy outcomes: current state of knowledge, challenges and potential implications for practice. Clin Endocrinol. 2018;88(6):761–9.
14. Palomba S, de Wilde MA, Falbo A, et al. Pregnancy complications in women with polycystic ovary syndrome. Hum Reprod Update. 2015;21(5):575–92.
15. Barrett ES, Hoeger KM, Sathyanarayana S, et al. Anogenital distance in newborn daughters of women with polycystic ovary syndrome indicates fetal testosterone exposure. J Dev Orig Health Dis. 2018;9(3):307–14.
16. Homburg R, Gudi A, Shah A, et al. A novel method to demonstrate that pregnant women with polycystic ovary syndrome hyper-expose their fetus to androgens as a possible stepping stone for the developmental theory of PCOS. A pilot study. Reprod Biol Endocrinol. 2017;15(1):61.
17. Risal S, Pei Y, Lu H, et al. Prenatal androgen exposure and transgenerational susceptibility to polycystic ovary syndrome. Nat Med. 2019;25(12):1894–904.
18. Piltonen TT, Giacobini P, Edvinsson Å, et al. Circulating antimüllerian hormone and steroid hormone levels remain high in pregnant women with polycystic ovary syndrome at term. Fertil Steril. 2019;111(3):588–96.
19. Tata B, Mimouni N, Barbotin AL, et al. Elevated prenatal anti-Müllerian hormone reprograms the fetus and induces polycystic ovary syndrome in adulthood. Nat Med. 2018;24(6):834–46.
20. Che X, Chen Z, Liu M, et al. Dietary interventions: a promising treatment for polycystic ovary syndrome. Ann Nutr Metab. 2021;77(6):313–23.
21. Barrea L, Arnone A, Annunziata G, et al. Adherence to the Mediterranean diet, dietary patterns and body composition in women with polycystic ovary syndrome (PCOS). Nutrients. 2019;11(10):2278.
22. Paoli A, Mancin L, Giacona MC, et al. Effects of a ketogenic diet in overweight women with polycystic ovary syndrome. J Transl Med. 2020;18(1).
23. O'Neill B, Raggi P. The ketogenic diet: pros and cons. Atherosclerosis. 2020;292:119–26.
24. Mavropoulos JC, Yancy WS, Hepburn J, et al. The effects of a low-carbohydrate, ketogenic diet on the polycystic ovary syndrome: a pilot study. Nutr Metab (Lond). 2005;2:35.
25. Ruan X, Li M, Mueck AO. Why does polycystic ovary syndrome (PCOS) need long-term management? Curr Pharm Des. 2018;24(39):4685–92.
26. Teede HJ, Misso ML, Costello MF, et al. Recommendations from the international evidence-based guideline for the assessment and management of polycystic ovary syndrome. Fertil Steril. 2018;110(3):364–79.
27. Szczuko M, Skowronek M, Zapałowska-Chwyć M, et al. Quantitative assessment of nutrition in patients with polycystic ovary syndrome (PCOS). Rocz Panstw Zakl Hig. 2016;67(4):419–26.

28. Baillargeon JP, Diamanti-Kandarakis E, Ostlund RJ, et al. Altered D-chiro-inositol urinary clearance in women with polycystic ovary syndrome. Diabetes Care. 2006;29(2):300–5.
29. Nordio M, Basciani S, Camajani E. The 40:1 myo-inositol/D-chiro-inositol plasma ratio is able to restore ovulation in PCOS patients: comparison with other ratios. Eur Rev Med Pharmacol Sci. 2019;23(12):5512–21.
30. Bevilacqua A, Dragotto J, Giuliani A, et al. Myo-inositol and D-chiro-inositol (40:1) reverse histological and functional features of polycystic ovary syndrome in a mouse model. J Cell Physiol. 2019;234(6):9387–98.
31. Chu W, Han Q, Xu J, et al. Metagenomic analysis identified microbiome alterations and pathological association between intestinal microbiota and polycystic ovary syndrome. Fertil Steril. 2020;113(6):1286–98.
32. Torres PJ, Siakowska M, Banaszewska B, et al. Gut microbial diversity in women with polycystic ovary syndrome correlates with hyperandrogenism. J Clin Endocrinol Metab. 2018;103(4):1502–11.
33. Qi X, Yun C, Sun L, et al. Gut microbiota-bile acid-interleukin-22 axis orchestrates polycystic ovary syndrome. Nat Med. 2019;25(8):1225–33.
34. Guarner F, Khan AG, Garisch J, et al. World Gastroenterology Organisation Global guidelines: probiotics and prebiotics October 2011. J Clin Gastroenterol. 2012;46(6):468–81.
35. Yurtdas G, Akdevelioglu Y. A new approach to polycystic ovary syndrome: the gut microbiota. J Am Coll Nutr. 2020;39(4):371–82.
36. Chudzicka-Strugała I, Kubiak A, Banaszewska B, et al. Effects of synbiotic supplementation and lifestyle modifications on women with polycystic ovary syndrome. J Clin Endocrinol Metab. 2021;106(9):2566–73.
37. Yurtdaş G, Akdevelioğlu Y. A new approach to polycystic ovary syndrome: the gut microbiota. J Am Coll Nutr. 2020;39(4):371–82.
38. Gholizadeh SS, Dehgan P, Mohammad-Alizadeh CS, et al. The effect of resistant dextrin as a prebiotic on metabolic parameters and androgen level in women with polycystic ovarian syndrome: a randomized, triple-blind, controlled, clinical trial. Eur J Nutr. 2019;58(2):629–40.
39. Shamasbi SG, Ghanbari-Homayi S, Mirghafourvand M. The effect of probiotics, prebiotics, and synbiotics on hormonal and inflammatory indices in women with polycystic ovary syndrome: a systematic review and meta-analysis. Eur J Nutr. 2020;59(2):433–50.
40. Karamali M, Eghbalpour S, Rajabi S, et al. Effects of probiotic supplementation on hormonal profiles, biomarkers of inflammation and oxidative stress in women with polycystic ovary syndrome: a randomized, double-blind, placebo-controlled trial. Arch Iran Med. 2018;21(1):1–7.
41. Wang FF, Wu Y, Zhu YH, et al. Pharmacologic therapy to induce weight loss in women who have obesity/overweight with polycystic ovary syndrome: a systematic review and network meta-analysis. Obes Rev. 2018;19(10):1424–45.
42. Ruan X, Song J, Gu M, et al. Effect of Diane-35, alone or in combination with orlistat or metformin in Chinese polycystic ovary syndrome patients. Arch Gynecol Obstet. 2018;297(6):1557–63.
43. Song J, Ruan X, Gu M, et al. Effect of orlistat or metformin in overweight and obese polycystic ovary syndrome patients with insulin resistance. Gynecol Endocrinol. 2018;34(5):413–7.
44. Yang PK, Hsu CY, Chen MJ, et al. The efficacy of 24-month metformin for improving menses, hormones, and metabolic profiles in polycystic ovary syndrome. J Clin Endocrinol Metab. 2018;103(3):890–9.
45. Huang X, Wang P, Tal R, et al. A systematic review and meta-analysis of metformin among patients with polycystic ovary syndrome undergoing assisted reproductive technology procedures. Int J Gynaecol Obstet. 2015;131(2):111–6.
46. Løvvik TS, Carlsen SM, Salvesen Ø, et al. Use of metformin to treat pregnant women with polycystic ovary syndrome (PregMet2): a randomised, double-blind, placebo-controlled trial. Lancet Diabetes Endocrinol. 2019;7(4):256–66.
47. Fornes R, Simin J, Nguyen MH, et al. Pregnancy, perinatal and childhood outcomes in women with and without polycystic ovary syndrome and metformin during pregnancy: a nationwide population-based study. Reprod Biol Endocrinol. 2022;20(1):30.

48. Hanem L, Stridsklev S, Júlíusson PB, et al. Metformin use in PCOS pregnancies increases the risk of offspring overweight at 4 years of age: follow-up of two RCTs. J Clin Endocrinol Metab. 2018;103(4):1612–21.
49. Hanem L, Salvesen Ø, Juliusson PB, et al. Intrauterine metformin exposure and offspring cardiometabolic risk factors (PedMet study): a 5-10 year follow-up of the PregMet randomised controlled trial. Lancet Child Adolesc Health. 2019;3(3):166–74.
50. Faure MC, Khoueiry R, Quanico J, et al. In utero exposure to metformin reduces the fertility of male offspring in adulthood. Front Endocrinol (Lausanne). 2021;12:750145.
51. Escobar-Morreale HF, Santacruz E, Luque-Ramírez M, et al. Prevalence of 'obesity-associated gonadal dysfunction' in severely obese men and women and its resolution after bariatric surgery: a systematic review and meta-analysis. Hum Reprod Update. 2017;23(4):390–408.
52. Daigle CR, Brethauer SA, Tu C, et al. Which postoperative complications matter most after bariatric surgery? Prioritizing quality improvement efforts to improve national outcomes. Surg Obes Relat Dis. 2018;14(5):652–7.
53. Jakobsen GS, Småstuen MC, Sandbu R, et al. Association of bariatric surgery vs medical obesity treatment with long-term medical complications and obesity-related comorbidities. JAMA. 2018;319(3):291–301.
54. Axelsson KF, Werling M, Eliasson B, et al. Fracture risk after gastric bypass surgery: a retrospective cohort study. J Bone Miner Res. 2018;33(12):2122–31.
55. Switzer NJ, Marcil G, Prasad S, et al. Long-term hypovitaminosis D and secondary hyperparathyroidism outcomes of the Roux-en-Y gastric bypass: a systematic review. Obes Rev. 2017;18(5):560–6.
56. Huhmann K. Menses requires energy: a review of how disordered eating, excessive exercise, and high stress lead to menstrual irregularities. Clin Ther. 2020;42(3):401–7.
57. Li Y, Ruan X, Wang H, et al. Comparing the risk of adverse pregnancy outcomes of Chinese patients with polycystic ovary syndrome with and without antiandrogenic pretreatment. Fertil Steril. 2018;109(4):720–7.

Ethical Issues in Fertility-Sparing Treatments in Gynecological Oncology

20

Simoncini Tommaso and Caretto Marta

Introduction

In 2020, there will be approximately 89,500 new cancer cases and 9270 cancer deaths in adolescents and young adults (AYAs) ages 15–39 years in the United States. Adolescents and young adults (AYAs) patients also have a high risk of long-term and late effects, including infertility, sexual dysfunction, heart problems, and future cancers. In AYAs, fertility is an important factor for good quality of life. In the case of cancer of the female reproductive tract, treatment can impair fertility, and therefore, AYAs may face the life-changing decision whether or not to undergo conservative, fertility-sparing cancer treatment. Each year, over 1,300,000 women are diagnosed with a gynecologic malignancy worldwide. Nearly 15% of these women are between 15 and 39 years of age. Nowadays, 80% of AYA cancer patients survive their disease because of the improved early detection and advancements in cancer treatment of many cancer types [1]. As a result, the focus of oncologic treatment has expanded from survival only toward quality of life after surviving cancer [2, 3]. However, fertility can be impaired by surgery or the gonadotoxic effects of chemotherapy and radiotherapy. Fertility is highly at risk in the case of treatment of malignancies of the female genital tract, especially cervical, ovarian, and endometrial cancer. Standard treatment for these cancer types often includes hysterectomy and bilateral salpingo-oophorectomy and, depending on stage, (adjuvant) therapy in the form of pelvic radiation or chemotherapy. Fertility-sparing surgery (FSS), in which ovaries, uterus, and sometimes cervix are (partially) preserved, is being offered in selected cases. Preservation of fertility should be discussed with premenopausal women with early-stage gynecologic cancer shortly after diagnosis

S. Tommaso (✉) · C. Marta
Division of Obstetrics and Gynecology, Department of Clinical and Experimental Medicine, University of Pisa, Pisa, Italy
e-mail: tommaso.simoncini@med.unipi.it

J. G. Schenker et al. (eds.), *Hot Topics in Human Reproduction*, Reproductive Medicine for Clinicians 3, https://doi.org/10.1007/978-3-031-24903-7_20

and, for women who desire to preserve fertility, during treatment planning. Criteria for patient selection for fertility-sparing therapy are not well defined; thus patients and providers must carefully discuss potential risks and benefits. In general, in carefully selected patients, survival outcomes do not appear to differ significantly between radical and fertility-sparing approaches. Women who undergo fertility-sparing therapies may experience a number of fertility and obstetric complications. Determining which women with gynecologic cancer are appropriate candidates for fertility-sparing treatments, assessing fertility potential, and helping women conceive after cancer treatment are best accomplished through multidisciplinary collaboration between gynecologic oncologists and fertility specialists [5]. However, the multidisciplinary team needs also to be updated about ethical consideration and balance of fertility preservation and cancer treatment.

Fertility-Sparing Surgery

The effect of cancer treatments on fertility and pregnancy outcomes is a distressing concern among the increasing population of reproductive-age women with gynecologic cancer [1]. The 2012–2016 Surveillance, Epidemiology, and End Results (SEER) statistics report 36.5% of cervical cancers, 6.5% of uterine cancers, and 7% of ovarian cancers were diagnosed in women <45 years old. Among all gynecologic cancers 44% of women with cervical cancer are diagnosed at an early stage, and almost 70% of endometrial cancers are diagnosed while still confined to the uterus. For ovarian cancer, about 14% are diagnosed with early-stage disease only. As women continue to delay childbearing, the number of young women with cancer who face fertility preservation decisions grows. Surgical interventions, gonadotoxic chemotherapy agents, and radiation therapy can have long-term detrimental effects on a woman's ability to conceive and/or successfully carry a pregnancy [6].

The Role of Fertility Specialists in Preserving Fertility

Reproductive physicians play important roles in helping to preserve the reproductive capacities of young cancer patients. First, they are involved in developing and using procedures to preserve gametes, embryos, and gonadal tissue before treatment. Second, fertility specialists will assist cancer survivors in using preserved gametes and tissue or in providing other assistance in reproduction [7].

Variations in type of cancer, time available until the onset of treatment, age, partner status, type and dosage of any chemotherapy and radiotherapy, and the risk of sterility with a given treatment regimen require that each case has its own treatment strategy. Consultation with the patient's oncologist is essential. Questions about the patient's health and prognosis also will arise when the patient is deciding later whether to reproduce. When a partner exists, he or she may be included in the discussion, but it is also advisable to discuss these issues with the patient individually.

Ethical Issue

Several ethical considerations should be taken in consideration when considering FSS and assisted reproduction [5]. In considering treatment options for cancer patients, *nonmaleficence* come first; the effects of both FSS and potential future pregnancy on recurrence and cancer outcomes, as well as the risk of pregnancy itself, must be considered. *Beneficence*, the second pillar of clinical medical ethics, requires consideration of the success of FSS at allowing cancer survivors to have children, ensuring FSS will provide the good that was intended. Third, patients' **autonomy** and consent require adequate patient education of traditional and fertility-sparing options, time for patients to consider these options, as well consideration for family influence. Lastly, since FSS often requires the use of assisted reproductive technology (ART) which has limited availability or can be cost-prohibitive, there is the issue of **justice** and patient access [8].

Nonmaleficence

The practical application of nonmaleficence is for the physician to weigh the benefits against burdens of all interventions and treatments, to avoid those that are inappropriately burdensome, and to choose the best course of action for the patient. This is particularly important and pertinent in difficult end-of-life care decisions on withholding and withdrawing life-sustaining treatment and medically administered nutrition and hydration and in pain and other symptoms' control [9].

Beneficence

The principle of beneficence is the obligation of physician to act for the benefit of the patient and supports a number of moral rules to protect and defend the right of others, prevent harm, remove conditions that will cause harm, help persons with disabilities, and rescue persons in danger. The principle calls for not just avoiding harm, but also to benefit patients and to promote their welfare [8]. While physicians' beneficence conforms to moral rules, and is altruistic, it is also true that in many instances it can be considered a payback for the debt to society for education (often subsidized by governments), ranks, and privileges and to the patients themselves (learning and research).

The Patient's Dilemma: Balancing Cancer and Fertility in Gynecologic Oncology

These first two principles can be traced back to the time of Hippocrates "to help and do no harm."

The health-care providers need to understand and discuss with patients about the true dilemma: balancing the cancer outcomes with the risk for fertility and pregnancy. The risk of pregnancy itself has not been shown to increase cancer recurrence for several tumors, including hormone-sensitive breast cancer, or to incur cancer-related harm to the offspring.

Fertility is highly at risk in the case of treatment of malignancies of the female genital tract, especially cervical, ovarian, and endometrial cancer.

Cervical Cancer

Of all cervical cancers, 20% is diagnosed in AYAs, making cervical cancer the most common gynecologic malignancy and the second cause of cancer-related death in these young women [1]. Fortunately, incidence and mortality are declining in some areas of the world due to population screening and human papilloma virus (HPV) vaccination.

However, more than 110,000 AYAs are diagnosed with cervical cancer, and over 31,000 AYAs still die of the disease each year worldwide [1]. (Radical) hysterectomy with or without pelvic lymphadenectomy is considered standard treatment for early-stage cervical cancer (FIGO 2018 IA1-IB2) in women who do not want to have children anymore. In women with a strong desire to preserve fertility, FSS options include conization or simple trachelectomy (cone or barrel-shaped excision of the cervix without surgery of the parametrium) or (vaginal or abdominal) radical trachelectomy (removal of the cervix, parametrium, and upper vaginal cuff), leaving the uterine body intact. These fertility-sparing procedures are increasingly being offered, as more and more studies suggest acceptable oncological outcomes comparable to radical hysterectomy [4]. However, the pregnancy rates especially after abdominal radical trachelectomy are disappointing [10]. The number of live births after abdominal radical trachelectomy in stage IB2 is only 9% [11]. Complications after FSS for cervical cancer such as cervical stenosis and Asherman syndrome—a disease characterized by scar tissue and adhesion formation within the uterus—can negatively impact fertility. Obstetrical complications after trachelectomy include miscarriage, preterm delivery, and preterm premature rupture of membranes. It is important to explain these risks to patients who elect fertility-sparing management and advise consultation with an obstetric specialist prior to conception [6]. Pregnancy rate after neoadjuvant chemotherapy (NACT) followed by conservative surgery is promising, but oncological safety is still unclear.

Ovarian Cancer

Ovarian cancer is the fourth most commonly diagnosed cancer in AYAs [1]. Epithelial ovarian cancer is the most common type, although non-epithelial ovarian cancer occurs more often in young women than in women over 40 years of age [2]. The incidence of ovarian cancer in AYAs represents 13% of all new diagnoses annually. The implication is that approximately 38,500 young women are diagnosed with ovarian cancer and that 10,000 of these women die from the consequences of the disease each year worldwide [1].

The standard management of clinical early-stage epithelial ovarian cancer is surgical staging, which includes hysterectomy, bilateral salpingo-oophorectomy, omentectomy, peritoneal washings, and biopsies with or without pelvic and para-aortic lymph node sampling. Depending on final stage and histology, platinum-based adjuvant chemotherapy can be proposed. During FSS, removal of the contralateral ovary and uterus are omitted. This conservative treatment is considered in AYAs with a strong desire to preserve fertility and with limited disease and no visible abnormalities during surgery. However, only observational, retrospective series comparing FSS and non-FSS in selected patients are available, and controversy about women with high-risk prognostic factors remains [12, 13]. Any surgery that involves removal of ovarian tissue may impact both immediate and long-term ovarian reserve, requiring some women to undergo ovarian stimulation with either oral agents or injectable gonadotropins in order to ovulate and achieve pregnancy.

Endometrial Cancer

Although only 4% of endometrial cancers occur in AYAs, this is still a global incidence of approximately 15,000 newly diagnosed women each year, of which 1600 young women die of the disease [1].

In endometrioid endometrial cancer, standard management involves total hysterectomy and bilateral salpingo-oophorectomy, leading to very high cure rates of 93% in low-risk disease [2]. The fertility-sparing alternative treatment includes hysteroscopic resection and/or curettage in combination with hormonal therapy with progestin. Complete remission rates of this fertility-sparing approach of 50–75% have been reported, demonstrating the clear concession to the high effectiveness of the standard treatment by hysterectomy [4, 6]. Strict follow-up with hysteroscopic evaluation and endometrial sampling is advised.

The level of evidence on oncological safety and chance of successful pregnancy after FSS in patients with a gynecological malignancy is low, because it is mainly based on retrospective case series including small numbers of patients and events. Moreover, follow-up is often short, and incidence of pregnancy and pregnancy outcome are incompletely reported. This complicates adequate counseling of AYAs with gynecological cancer, and the wish to preserve fertility consequently hampers these women to make the life-changing choice they are facing. Despite successful hormonal treatment of endometrial cancer, many young women still face subfertility due to underlying metabolic disorders, such as polycystic ovary syndrome. These women often have a reduced rate of conception and live birth, possibly due to factors that also contributed to the development of endometrial cancer (chronic anovulation and/or obesity). Further, their endometrial pathology both before and after treatment may not promote embryonic implantation and development [6]. It is recommended that ART be started as soon as response to hormonal therapy is achieved, to maximize pregnancy success and minimize time prior to definitive surgery with hysterectomy and thereby minimize the risk of relapse. Ultimately, immediate ART helps to avoid prolonged, unopposed estrogen stimulation, which could also result in relapse and disease progression [6].

Oncologic Safety of Assisted Reproductive Technology in Women with a History of Gynecologic Cancer

An important concern for women with a history of gynecologic cancer considering assisted reproduction is the impact of ART on the risk of cancer recurrence.

Several factors complicate research exploring cancer development or recurrence after ART. Risk factors for infertility and gynecologic cancers often overlap, making it difficult to assess causation in the relationship between assisted reproduction and cancer development. The field of ART continues to evolve at a fast pace, making evaluation of long-term outcomes a challenge as well [14]. Quantification of risk should be individualized, and further prospective studies are necessary to better ensure the safety pro le of assisted reproduction in women with a history of gynecologic cancer.

Consent/Autonomy

This ethical principle was affirmed in a court decision by Justice Cardozo in 1914 with the epigrammatic dictum, "Every human being of adult years and sound mind has a right to determine what shall be done with his own body" [15]. Autonomy, as is true for all four principles, needs to be weighed against competing moral principles and in some instances may be overridden. The principle of autonomy does not extend to persons who lack the capacity (competence) to act autonomously; examples include infants and children and incompetence due to developmental, mental, or physical disorder. Health-care institutions and state governments in the United States have policies and procedures to assess incompetence. However, a rigid distinction between incapacity to make health-care decisions (assessed by health professionals) and incompetence (determined by court of law) is not of practical use, as a clinician's determination of a patient's lack of decision-making capacity based on physical or mental disorder has the same practical consequences as a legal determination of incompetence [9].

Resistance to the principle of patient autonomy and its derivatives (informed consent, truth-telling) in non-western cultures is not unexpected. In countries with ancient civilizations, rooted beliefs, and traditions, the practice of paternalism (this term will be used in this article, as it is well-entrenched in ethics literature, although parentalism is the proper term) by physicians emanates mostly from beneficence. However, culture (a composite of the customary beliefs, social forms, and material traits of a racial, religious, or social group) is not static and autonomous and changes with other trends over passing years. It is presumptuous to assume that the patterns and roles in physician-patient relationships that have been in place for a half a century and more still hold true. Therefore, a critical examination of paternalistic medical practice is needed for reasons that include technological and economic progress, improved educational and socioeconomic status of the populace, globalization, and societal movement toward emphasis on the patient as an individual, than as a member of a group. This needed examination can be accomplished by research that includes well-structured surveys on demographics, patient preferences on informed consent, truth-telling, and role in decision-making. Respecting the principle of autonomy obliges the physician to disclose medical information and treatment options that are necessary for the patient to exercise self-determination and supports informed consent, truth-telling, and confidentiality [9].

Informed Consent

The requirements of an informed consent for a medical or surgical procedure, or for research, are that the patient or subject:

- Must be competent to understand and decide
- Receives a full disclosure
- Comprehends the disclosure
- Acts voluntarily
- Consents to the proposed action

The universal applicability of these requirements, rooted and developed in western culture, has met with some resistance and a suggestion to craft a set of requirements that accommodate the cultural mores of other countries [16].

As competence is the first of the requirements for informed consent, one should know how to detect incompetence. Standards (used singly or in combination) that are generally accepted for determining incompetence are based on the patient's inability to state a preference or choice, inability to understand one's situation and its consequences, and inability to reason through a consequential life decision.

In a previously autonomous, but presently incompetent patient, his/her previously expressed preferences (i.e., prior autonomous judgments) are to be respected [17]. Incompetent (non-autonomous) patients and previously competent (autonomous) but presently incompetent patients would need a surrogate decision-maker. In a non-autonomous patient, the surrogate can use either a substituted judgment standard (i.e., what the patient would wish in this circumstance and not what the surrogate would wish) or a best interests standard (i.e., what would bring the highest benefit to the patient by weighing risks and benefits). Snyder and Sulmasy [18], in their thoughtful article, provide a practical and useful option when the surrogate is uncertain of the patient's preference(s) or when patient's preferences have not kept abreast of scientific advances. They suggest the surrogate use "substituted interests," that is, the patient's authentic values and interests, to base the decision.

In gynecologic oncology, the real difficult issue about the consent form is the situation in which the final operative decision occurs when the women are under anesthesia. For a young woman with ovarian mass, the ultimate decision will occur after the macroscopic evaluation of the abdominal cavity and after the frozen analysis of the pathology. Most of the consent process must thus happen theoretically, prior to the final pathology result. Despite a comprehensive educational oncology pathway, patients retain little of the informed consent discussion. There is a dichotomy between the outcomes that surgeons and patients' value most: consent for young patients in these situations may require multiple discussions and a good knowledge of the patient's priorities so the surgeons can make the appropriate decision.

Justice

Justice is generally interpreted as fair, equitable, and appropriate treatment of persons. Of the several categories of justice, the one that is most pertinent to clinical ethics is distributive justice. Distributive justice refers to the fair, equitable, and appropriate distribution of health-care resources determined by justified norms that structure the terms of social cooperation [19]. How can this be accomplished? There are different valid principles of distributive justice. These are distribution to each person:

- An equal share
- According to need
- According to effort
- According to contribution
- According to merit
- According to free-market exchanges

Each principle is not exclusive, and can be, and are often combined in application. It is easy to see the difficulty in choosing, balancing, and refining these principles to form a coherent and workable solution to distribute medical resources.

In the setting of AYA patients with cancer, justice includes equitable access to FSS, which has been shown to vary. Many women need ART to become pregnant after FSS, which can be prohibitively expensive for many couples. Lastly, even the option of adoption may be limited for cancer survivors. Several fertility treatments are not covered by insurance and are often expensive. Women with lower income may not be able to benefit from FSS, despite incurring the risk (if there are potential differences in outcome or need for closer surveillance).

On the other hand, child adoption can be a viable option for many women, who lost their fertility due to cancer or cancer treatment. However, some adoption agencies may require a certain cancer-free interval prior to considering a parent who is a cancer survivor, and not all agencies can offer cancer survivors protection against discrimination by the parents placing their child up for adoption.

Conclusion

Oncologists should be encouraged to refer patients to a reproductive endocrinologist early in the planning of treatment. When damage to reproductive organs due to gonadotoxic treatment is unavoidable, health-care providers should inform patients of options to preserve fertility. Counseling by a qualified mental-health professional and genetic counselor, when appropriate, also should be offered: a collaborative multidisciplinary team approach is encouraged.

Ethics is an inherent and inseparable part of clinical medicine as the physician has an ethical obligation to benefit the patient, to avoid or minimize harm, and to respect the values and preferences of the patient.

Appropriate candidates for fertility-sparing treatment need to be identified (Table 20.1) according to ethic criteria. The concept of nonmaleficence can be applied primarily to oncologic and surgical outcomes: FSS should not be recommended if it confers a worsening of oncologic outcomes. The potential for high-risk pregnancy should also be discussed. In terms of beneficence, women must have a reasonable expectation that they could conceive based on clinical and social factors. The consent process, a reflection of patient autonomy, needs to consider the balance between the clinical assessment for both the cancer and fertility. Lastly, it must be addressed that many patients cannot afford the reproductive technology that they would need for reproduction even after FSS. Clinicians should inform patients receiving potentially gonadotoxic therapies about options for fertility preservation and future reproduction prior to the initiation of such treatment.

Table 20.1 Criteria for candidates for fertility-sparing surgery

Criteria for fertility-sparing surgery
Oncologic factors (nonmaleficence)
• Equivalent oncologic outcomes
• Good prognosis
• Ability to comply with close surveillance
• Willingness to undergo definitive surgery in case of recurrence
Reproductive factors (beneficence)
• Desire future fertility
• Young age
• Reasonable probability of pregnancy based on clinical fertility evaluation
Patient consent (autonomy)
• Understanding of risk and potential benefits
• Understanding of minor deviation from standard of care versus experimental treatments
• Plan for intraoperative decision-making
• Incorporates cultural family influence regarding fertility, if appropriate
Access to cure (justice)
• Discuss FSS with all premenopausal women, explain whether and why they are a candidate
• Ensure that the patient understands the need for assisted reproductive technology and its financial implication, if appropriate.

Acknowledgments
Funding: No external funding has been received for the preparation of this chapter.

Disclosure: No disclosure to declare.

References

1. World Health Organization, International Agency for Research on Cancer (IARC), Global Cancer Observatory (GCO). https://gco.iarc.fr. Accessed 17 Nov 2020.
2. Moser EC, Meunier F. Cancer survivorship: a positive side-effect of more successful cancer treatment. EJC Suppl. 2014;12(1):1–4. https://doi.org/10.1016/j.ejcsup.2014.03.001.
3. Duffy C, Allen S. Medical and psychosocial aspects of fertility after cancer. Cancer J. 2009;15(1):27–33. https://doi.org/10.1097/PPO.0b013e3181976602.
4. Prodromidou A, Iavazzo C, Fotiou A, Psomiadou V, Douligeris A, Vorgias G, Kalinoglou N. Short- and long term outcomes after abdominal radical trachelectomy versus radical hysterectomy for early stage cervical cancer: a systematic review of the literature and meta-analysis. Arch Gynecol Obstet. 2019;300(1):25–31. https://doi.org/10.1007/s00404-019-05176-y.
5. Schuurman T, Zilver S, Samuels S, Schats W, Amant F, van Trommel N, Lok C. Fertility-sparing surgery in gynecologic cancer: a systematic review. Cancers (Basel). 2021;13(5):1008. https://doi.org/10.3390/cancers13051008.
6. Floyd JL, Campbell S, Rauh-Hain JA, Woodard T. Fertility preservation in women with early-stage gynecologic cancer: optimizing oncologic and reproductive outcomes. Int J Gynecol Cancer. 2021;31(3):345–51. https://doi.org/10.1136/ijgc-2020-001328.
7. Ethics Committee of the American Society for Reproductive Medicine. Electronic address: ASRM@asrm.org. Fertility preservation and reproduction in patients facing gonadotoxic therapies: an Ethics Committee opinion. Fertil Steril. 2018;110(3):380–6. https://doi.org/10.1016/j.fertnstert.2018.05.034.
8. Varkey B. Principles of clinical ethics and their application to practice. Med Princ Pract. 2021;30(1):17–28. https://doi.org/10.1159/000509119. Epub 2020 Jun 4
9. Beauchamp TL, Childress JF. Principles of biomedical ethics. New York, NY: Oxford University Press; 2009. p. 162–4.
10. Bentivegna E, Maulard A, Pautier P, Chargari C, Gouy S, Morice P. Fertility results and pregnancy outcomes after conservative treatment of cervical cancer: a systematic review of the literature. Fertil Steril. 2016;106(5):1195–1211.e5. https://doi.org/10.1016/j.fertnstert.2016.06.032.
11. Cao DY, Yang JX, Wu XH, Chen YL, Li L, Liu KJ, Cui MH, Xie X, Wu YM, Kong BH, Zhu GH, Xiang Y, Lang JH, Shen K, China Gynecologic Oncology Group. Comparisons of vaginal and abdominal radical trachelectomy for early-stage cervical cancer: preliminary results of a multi-center research in China. Br J Cancer. 2013;109(11):2778–82. https://doi.org/10.1038/bjc.2013.656.
12. Bentivegna E, Gouy S, Maulard A, Pautier P, Leary A, Colombo N, Morice P. Fertility-sparing surgery in epithelial ovarian cancer: a systematic review of oncological issues. Ann Oncol. 2016;27(11):1994–2004. https://doi.org/10.1093/annonc/mdw311.
13. Liu D, Cai J, Gao A, Wang Z, Cai L. Fertility sparing surgery vs radical surgery for epithelial ovarian cancer: a meta-analysis of overall survival and disease-free survival. BMC Cancer. 2020;20(1):320. https://doi.org/10.1186/s12885-020-06828-y.
14. Levine RJ. Informed consent: some challenges to the universal validity of the western model. In: Vaughn L, editor. Bioethics: principles, issues and cases. New York, NY: Oxford University Press; 2010. p. 183–8.

15. Luke B, Brown MB, Spector LG, Missmer SA, Leach RE, Williams M, Koch L, Smith Y, Stern JE, Ball GD, Schymura MJ. Cancer in women after assisted reproductive technology. Fertil Steril. 2015;104(5):1218–26. https://doi.org/10.1016/j.fertnstert.2015.07.1135.
16. Cardozo B. Basic right to consent to medical care—Schlendorff vs the Society of the New York Hospital, 211 NY 125 105 NE 92 1914 LEXUS 1028 (1914).
17. Davis JK. The concept of precedent autonomy. Bioethics. 2002;16(2):114–33.
18. Sulmasy DP, Snyder L. Substituted interests and best judgments: an integrated model of surrogate decision making. JAMA. 2010;304(17):1946–7. https://doi.org/10.1001/jama.2010.1595.
19. Fleishacker S. A short history of distributive justice. Cambridge, MA: Harvard University Press; 2005.

GPSR Compliance

The European Union's (EU) General Product Safety Regulation (GPSR) is a set of rules that requires consumer products to be safe and our obligations to ensure this.

If you have any concerns about our products, you can contact us on ProductSafety@springernature.com

In case Publisher is established outside the EU, the EU authorized representative is:

Springer Nature Customer Service Center GmbH
Europaplatz 3
69115 Heidelberg, Germany

Batch number: 10732090

Printed by Printforce, the Netherlands